Quantitative Methods in the Humanities and Social Sciences

Series Editors

Thomas DeFanti, University of California San Diego, La Jolla, CA, USA
Anthony Grafton, Princeton University, Princeton, NJ, USA
Thomas E. Levy, University of California San Diego, La Jolla, CA, USA
Lev Manovich, The Graduate Center, CUNY, New York, NY, USA
Alyn Rockwood, KAUST, Boulder, CO, USA

More information about this series at http://www.springer.com/series/11748

Alan Shepherd

Let's Calculate Bach

Applying Information Theory and Statistics to Numbers in Music

Alan Shepherd
Dierdorf, Germany

ISSN 2199-0956 ISSN 2199-0964 (electronic)
Quantitative Methods in the Humanities and Social Sciences
ISBN 978-3-030-63768-2 ISBN 978-3-030-63769-9 (eBook)
https://doi.org/10.1007/978-3-030-63769-9

Mathematics Subject Classification: 94A15, 62P99, 65C05

© Springer Nature Switzerland AG 2021
This work is subject to copyright. All rights are reserved by the Publisher, whether the whole or part of the material is concerned, specifically the rights of translation, reprinting, reuse of illustrations, recitation, broadcasting, reproduction on microfilms or in any other physical way, and transmission or information storage and retrieval, electronic adaptation, computer software, or by similar or dissimilar methodology now known or hereafter developed.
The use of general descriptive names, registered names, trademarks, service marks, etc. in this publication does not imply, even in the absence of a specific statement, that such names are exempt from the relevant protective laws and regulations and therefore free for general use.
The publisher, the authors and the editors are safe to assume that the advice and information in this book are believed to be true and accurate at the date of publication. Neither the publisher nor the authors or the editors give a warranty, expressed or implied, with respect to the material contained herein or for any errors or omissions that may have been made. The publisher remains neutral with regard to jurisdictional claims in published maps and institutional affiliations.

This Springer imprint is published by the registered company Springer Nature Switzerland AG
The registered company address is: Gewerbestrasse 11, 6330 Cham, Switzerland

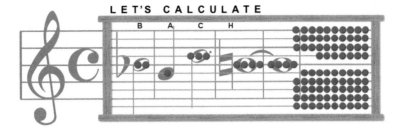

Illustration: © by the author.
In German the note B♭ is called B and B is called H, hence the name BACH gives the melody shown. The note values in quarter notes are 2, 1, 3, 8 which is BACH in the number alphabet with A = 1, B = 2, etc.

Calculemus.
Lasst uns rechnen.
Let us calculate.

Gottfried Wilhelm Leibniz (1646–1716)

Foreword

There is always one statement in a book that the author regrets almost as soon as it is published. In my case, it appears on page 140 of Bach's Numbers and reads: "The probability of six terms between 272 and 524 falling randomly into a perfect double 2:1 proportion is minimal. Bach must have planned it". Combining compositional intention with probability, it was a provocative invitation to statisticians. My regret was short-lived, though, as it quickly led to a fruitful collaboration with Alan Shepherd and to this book.

In October 2016, I received a short email from Alan inquiring about the 2017 Bach Network Dialogue Meeting, which included the passing comment: "I am very interested in your research on Bach's numbers and have your two books. I may even be able to contribute some thoughts". As Bach's Numbers: Compositional Proportion and Significance had only just been published, I realised this must be a serious reader. Two months later, Alan sent an abstract of the topic he was hoping to present at the Dialogue Meeting: "This paper applies some mathematics and information theory to map out the possibilities of encoding information in musical scores and puts this in perspective with some common biases and fallacies to which most humans are prone when evaluating probabilities such as how likely or improbable an observation is". We set up a FaceTime call.

Over the years, my research into the numerical aspects of Bach's compositional process has had a mixed reception, with equal measures of outright dismissal, enthusiastic acceptance and lukewarm scepticism. Critical scepticism was my default setting as a musical analyst. My aim was to discover what evidence, if any, there was behind the popular craze of interpreting numbers in the structures of Bach's scores. To do this systematically, I needed to know which number methods Bach could have known and which could have been applied to musical composition. My research turned up the origins of number alphabets, which Bach would have known, and an explanation of how and when the fashion for gematria became associated with Bach's music (Tatlow 1991). Next I had to establish a clear analytical method that would, as far as possible, uncover evidence of Bach's working method. Rather than using established analytical methods, I chose to use tools and concepts that Bach himself would recognise, as they promised to take me closer to Bach's thought processes. I scoured treatises that he owned or could have read for evidence of how he might have ordered his compositions numerically, if at all. I used pencil, paper and simple arithmetic as I looked at the structures of his scores, in their early, revised and final forms. This took me through an exploration of whether or not he could have used the golden section or the Fibonacci series (Tatlow 2006), to the discovery in 2004 of what seemed to be an extraordinary one-off use of the proportions 1:2 in several layers within the Six Solos for Violin (BWV 1001–1006). Later came the unforgettable moment when I realised that it was only in the final revisions that the bar totals were whole round numbers divided by clear large-scale layers of proportion. Until that point, I had stood safely on the side of the sceptics, but when I tentatively presented the first results in December 2006 a colleague challenged me to cross the line. The first iteration of the theory of proportional parallelism was published soon after (Tatlow 2007). Like it or not, I found myself presenting new evidence to suggest that Bach had deliberately used numbers to create proportional order in his compositions. Tatlow (2007) held me to a theory with a three-part test that I continued to refine as I criticised, wrote and rewrote what would become Bach's Numbers (Tatlow 2015). As part of the test, I asked a colleague in 2010 to run a statistical exploration of my results for the Six Solos for Violin to see the probability that these patterns were intentional on Bach's part. He concluded that the "null hypothesis" gave no reason to believe that my hypothesis was true. This ran counter to a gut instinct based on both my growing understanding of how Bach's contemporaries thought about music and the numerous examples in Bach's music with layers of large-scale double and sometimes triple 1:2 or 1:1 proportion. Was the negative result because my colleague had not factored in sufficient data? Was it that statistical techniques were unsuitable for historical studies? When Alan and I had our first conversation 6 years later these considerations were uppermost in my mind.

In Bach's Numbers, I wanted to present the evidence transparently so that scholars from all disciplines, with negative or positive expectations, could read and weigh up the evidence, test the theory for themselves, accept what rang true and reject and replace what did not. In Part I—Foundations—I present the strands of historical evidence on compositional process, parallel techniques and beliefs about proportions. In Part II—Demonstrations—I give evidence of the compositional history with tables of the final totals and their layers of 1:1 and 1:2 proportions for each of Bach's published works. Some tables are more convincing than others, and I like some of them more than others. It seemed a fairer and more honest test of the theory to include the demonstrations that I found less convincing. It is clear from the variety of results that Bach had principles of ordering, but that there was no rigid code to which he adhered.

I discovered that Bach's works were layered with numerous proportioned units nesting within each other, which I called parallel proportions. Combined with the historical evidence, it was the ingenuity of these layers that persuaded me they had been planned. A single 1:1 or 1:2 proportion formed by the number of bars, the number of movements or works or by their symmetrical arrangement could easily be coincidental. A double 1:1 or 1:2 proportion, formed by the number of bars, and the number of movements or works, and their symmetrical arrangement seemed more intentional, particularly when subdivided into a further 1:1 or 1:2 proportion, and even more so when the same three elements combined to form a triple 1:1 or 1:2 proportion. The case for intentionality became even stronger when these proportions were further combined with double or triple 1:1 or 1:2 proportions at different structural levels, especially when these two were often layered with two or more layers of double or triple 1:1 or 1:2 proportion.

In the structure of Bach's Six Solos for Violin, for example, the largest-scale double 1:1 proportion is between the 6 Solos:6 Sonatas each with 2400:2400 bars (Tatlow 2015, 133–158). The next large-scale double 2:1 proportion is within the Six Solos, where 4:2 Solos are proportioned as 1600:800 bars. This is parallel to another large-scale double 2:1 proportion within the Six Violin Sonatas, with 4:2 Sonatas in 1600:800 bars.

At a smaller scale, single 1:1 and 1:2 proportions are found in the four movements of the G minor Sonata (BWV 1001), with 136:136 bars and 136:272 bars (including repeats). This is parallel to yet another smaller-scale single 2:1 proportion in the four movements of the B minor Partita (BWV 1002), both 272:136 bars and 544:272 bars with repeats. As the proportions are formed with units of 136 bars, there are several additional layers of proportion between the G minor Sonata and the B minor Partita.

These results were supported by evidence of compositional adjustments that Bach made as the work evolved. Combined with the many good theological and aesthetic reasons that Bach and his contemporaries had for creating perfectly executed symmetry, 1:1 and 1:2 proportions, it felt acceptable in this instance to claim intentionality. In 2016, Alan Shepherd set out with an open mind to test my hypothesis, initially developing the Proportional Parallel Explorer program to find simple proportions and over time refining it to isolate simultaneous layering of proportions with symmetrical layouts. Of interest to us both were to see how far statistical and computer-based methods could shed light on whether Bach intended the results I had discovered, whether it was "analytical coincidence or Bach's design", as I asked in 2007 (Tatlow 2007).

It is fairly alarming to have one's work scrutinised in minute detail. It is also a great honour. When Alan Shepherd and I first spoke in 2016, we had no idea of what would result. For me, it has meant gaining a colleague and friend, burying any regrets about that sentence (see 10.4) and realising anew that good old-fashioned gut instinct still has a place in scholarship. I am grateful to Alan for his patience and humility, and delighted that with this book he has succeeded in pushing the boundaries of musicological methods, to narrow the gap at the intersection of compositional intentionality and numerical construction.

Danderyd, Sweden
July 2020

Ruth Tatlow

Preface

My interest in music started when I was about sixteen years old, when like most youths I learned to strum chords on the guitar. However, not being able to sing very well, there seemed little point in just playing chord sequences, so I changed to classical guitar and then piano and organ.

In my student days in the early 1970s studying electrical and electronic engineering at Brunel University, UK, my free-time interest centred on baroque music, especially J. S. Bach. I had had my first piano and organ lessons in my last two years at school and was able to continue these for some of the time at university. I used to go to Fenton House in London, which housed a collection of period instruments, and I was allowed to play the harpsichords there, including one actually played by Händel. I built a clavichord from a kit (Fig. 1), but this was not very successful. The university was also conveniently situated for attending concerts in London. Not being able to afford to buy many scores, I borrowed them from the library and copied them out, much as was done in the eighteenth century. Listening was mainly from the radio and later with a reel-to-reel tape recorder for which I built a stereo decoder for my portable monophonic radio. The combined interest in Bach and electronics was fed by Wendy Carlos' (or Walter Carlos at that time) recording "Switched on Bach" of Bach's music played on the Moog synthesiser. On a student exchange with Germany,

Fig. 1 Clavichord built from a kit. *Photograph* © the author

I happened to read a book in the local library which must have been (Smend 1950) and this piqued my interest in the subject of numerology.

This wider range of interests came up in my first job interview. I was asked why I only had a second class honours degree and not a first? I explained that I perhaps could have obtained a first if I had done nothing else but the exam subjects, but I also pursued interests and took courses in music, photography, German and read some English and German literature. Luckily, the owner of the company accepted this, and I got the job, which resulted in my move to Germany.

But pursuit of musical interests took a back seat to my family and my working life in programming real-time operating systems and industrial process control systems, and then in IT service management and IT governance. In the meantime, much more has been written about numbers in Bach, and the use of personal computers has opened up new avenues of enquiry, which I have not seen pursued in the literature so far. During my final working years, I was investigating the application of risk management to IT and learned some things about statistics, the pitfalls of dealing with probabilities and cognitive biases, particularly from Hubbard (2009, 2014) and Fenton. In fact, I learned of these books from my brother Dr. Keith Shepherd, who works as a leading scientist in a completely unrelated field of agricultural research.

Now, in retirement, I decided to apply my knowledge of information theory and technology, statistics and computer programming to numerical studies in music. What I thought might be a short paper has expanded into a book and a computer program.

I hope that this combination of information theory, computer technology, probability and statistics, cognitive psychology and music, supplemented by a computer program provided to assist in exploring certain aspects, can add some new perspectives and provide additional support for future musicological research.

Dierdorf, Germany   Alan Shepherd

Acknowledgements

My first debt of gratitude is owed to Dr. Ruth Tatlow, first for her books (Tatlow 1991, 2015) which inspired me to start this work, and then for her encouragement in writing it up for publication, her openness to the statistical approach, as well as being the first user of the program described in Chap. 8. She also provided the impulse to include something on proportions in architecture and put me right on several musicological points.

My wife Christine was not expecting me to start a large project after I retired and was very patient as the work dragged on.

My daughter Elisabeth proofread the draft, spotted many typing errors and was a good tester of the readability.

Much of my research was carried out at the Bach Archive and Library in Leipzig, so many thanks to the very helpful and knowledgeable staff there.

The editors and reviewers at my publisher, Springer-Verlag, firstly for accepting the book at all, and especially Veronika Rosteck who was very helpful to me as a first-time author.

Some of the anonymous reviews of the initial draft sent to Springer were very thorough and extremely useful to set me off on the right path—you know who you are, so please accept my gratitude.

Obtaining permissions to use images or take my own photographs was a varied experience with immediate positive responses, long delays and demands for exorbitant fees, extended quests to find and contact the owners, and rewarding exchanges with some very interesting people. For details on Creative Commons Licenses, see (Internet-CC-Licenses).

The computer program Proportional Parallelism Explorer was written in Java® (Internet1) and initially developed using the BlueJay (Internet2) development environment (while learning Java) and then Eclipse (Internet3). The program is versioned and distributed on GitHub (Internet4).

The graphics from the program's output were produced with the R statistics package (Internet5). The Bayesian networks were created with the AgenaRisk program, initially with the Lite version provided with Fenton and then with a full version kindly loaned by Norman Fenton. Norman also reviewed and corrected my example Bayesian network.

Musical notation in the frontispiece and in the text is in the "Bach" musicological font by Yo Tomita (Internet6).

Contents

1	**Introduction**		1
	1.1 The Science of Musicology		1
	1.2 Numerology and Bach		2
	1.3 About This Book		3
2	**An Information Theory Approach**		7
	2.1 Information and Communication		7
	2.2 Measuring Information—The Bit		8
	2.3 The Bit as Binary Digit		9
	2.4 Signal, Noise, Redundancy and Encoding		9
	2.5 Messages and Symbols		11
	2.6 Throughput and Protocols		13
	2.7 Gematria as Hash Coding		13
	2.8 An Unambiguous Coding		15
	2.9 Codings and References		16
3	**Some Possible Codings in Music**		19
	3.1 Preamble		19
	3.2 Number of Bars		20
	3.3 Notes		21
	3.4 Intervals		22
	3.5 Note Lengths		22
	3.6 Number of Notes		23
	3.7 Number of Pieces, Movements or Sections		24
	3.8 Sum of the G-Values of Notes		25
	3.9 Key Signature		25
	3.10 Accidentals		27
	3.11 Occurrences of Words		27
	3.12 Rests		27
	3.13 Time Signature		28
	3.14 Figured Bass		29
	3.15 Entries of a Theme		29
	3.16 Other Possibilities		29

		3.16.1	Acrostics	29
		3.16.2	More Subtle Ways	30
	3.17	Beyond Bach		31
		3.17.1	BWV Numbers	31
		3.17.2	Frequencies	31
		3.17.3	Morse Code	31
		3.17.4	Colours and Shapes	32
		3.17.5	Other Puzzles	33
	3.18	Combined Codings		34
	3.19	A Cryptographic Example		34
	3.20	Summary		35
	3.21	The Real Coding		36
	3.22	Notes for Researchers		38
4	**Ambiguity in Decoding**			39
	4.1	Preamble		39
	4.2	Sources		40
	4.3	Modern Dictionary		40
		4.3.1	Method	40
		4.3.2	Modern Dictionary with Latin Natural Coding	41
		4.3.3	Modern Dictionary with Latin Milesian and Trigonal Coding	42
	4.4	Historic Sources		44
		4.4.1	Luther Bible	44
		4.4.2	Cantata Texts	45
		4.4.3	Combining Historic Sources	45
	4.5	Summary		48
	4.6	Notes for Researchers		49
5	**Multiple Words and Partitioning**			51
	5.1	Partitioning and Permutations		51
	5.2	Partitioning G-Values		52
	5.3	Composers' Names		55
	5.4	Notes for Researchers		55
6	**Score Analysis**			57
	6.1	The Method		57
	6.2	Counting Bars		60
	6.3	Statistics		65
	6.4	Further Applications		67
	6.5	Summary		70
	6.6	Other Representations and Tools		71
	6.7	Notes for Researchers		72

7	**Statistical Methods**		73
	7.1	Preamble	73
	7.2	Probability and Distributions	74
	7.3	Hypothesis Testing and Significance	77
	7.4	Confidence Interval	79
	7.5	Monte Carlo Simulation	80
	7.6	Bayes Theorem	82
	7.7	Notes for Researchers	85
8	**Exploring Proportions**		87
	8.1	Preamble	87
	8.2	Simple Proportions and Terminology	93
		8.2.1 Sets and Pieces	93
		8.2.2 Proportion	94
		8.2.3 Combinations	95
		8.2.4 Solutions, Targets, Opposites and Complements	97
		8.2.5 Symmetries, Signatures and Patterns	98
		8.2.6 Binary Signatures	99
	8.3	Layers of Proportion	100
	8.4	Summary of Terms	102
	8.5	The Proportional Parallelism Explorer Program	103
		8.5.1 Solution Search	103
		8.5.2 Solutions Search Through Layers	107
		8.5.3 Pattern Matching	107
		8.5.4 Pattern Matching in Layers	110
		8.5.5 Colour Coding for Visual Pattern Recognition	111
		8.5.6 Monte Carlo Simulation	111
9	**Applying the Methods to the Well Tempered Clavier Book 1 BWV 846–869**		117
	9.1	Preamble	117
	9.2	Solutions	117
	9.3	Probability	118
	9.4	Monte Carlo Simulation	123
	9.5	Hypothesis Testing and Significance	127
	9.6	Bayes Theorem	128
	9.7	Patterns	133
	9.8	Preludes and Fugues Separately	141
	9.9	Ariadne Musica	149
10	**Consolidated Observations**		153
	10.1	Preamble	153
	10.2	The Effect of the Number of Pieces	153
	10.3	Works Which Could Have More Than One Layer	167
	10.4	Probability	167

10.5	Types of Distribution	172
10.6	Real Works Versus Single-Layer Simulations	174
10.7	Accuracy	179
10.8	Proportions and Other Structures	181
10.9	Proportions in Durations	182
10.10	Works with No Proportions	185
10.11	The Impossible Proportions	185
10.12	Reverse Engineering and the Art of Fugue BWV 1080	186
10.13	Combining Works	190
10.14	Summary of Main Statistics	196
10.15	Notes for Researchers	203

11 Magic Squares 205
    11.1 The Dieben Rectangle 205
        11.1.1 Deriving Proportions from Dieben's Rectangle 209
    11.2 Use by Modern Composers 212

12 Psychological Fallacies 213
    12.1 Preamble 213
    12.2 Story Bias or Narrative Fallacy 213
    12.3 Confirmation Bias 214
    12.4 Neglect of Probability 214
    12.5 Halo Effect 215
    12.6 Conjunction Fallacy 216
    12.7 Notes for Researchers 216

13 Bach, Science and Technology 217
    13.1 Mathematics and Philosophy in Bach's Time 217
    13.2 Models 220
    13.3 Computers 226
    13.4 Artificial Intelligence 228
    13.5 Quantz and Hi-Fi 230
    13.6 Summary 231

14 Conclusion 233

Appendix A: More Parallel Proportion Results 235

Appendix B: Proportional Parallelism Explorer Program User Manual 283

Appendix C: Tabular History 319

Appendix D: Alphabet Tables 323

Appendix E: Interval Proportions 325

Appendix F: Excel Functions 327
Literature 331
General Index 339
Index of Names 349

Abbreviations

AI	Artificial Intelligence
AMB	Anna Magdalena Bach
BD	Bach Dokumente (Bach Documents)
BWV	Bach Werke Verzeichnis (catalogue of J. S. Bach's works)
CD	Compact Disc
CIA	Central Intelligence Agency
CSV	Comma-Separated Variable
CÜ	Clavierübung
DARMS	Digital Alternate Representation of Musical Scores
DNA	Deoxyribonucleic Acid
HTML	HyperText Markup Language
ICT	Information and Communications Technology
IO	Input–Output
IT	Information Technology
JPEG	Joint Photographic Experts Group
JRE	Java Runtime Environment
LINUX	Linus (Torvalds') UNIX
MC	Monte Carlo
MEI	Music Encoding Initiative
MIDI	Musical Instruments Digital Interface
MP3	Moving Picture Experts Group Layer-3 Audio
NA	Not Applicable
NATO	North Atlantic Treaty Organization
NP	Nondeterministic Polynomial Time (NP-Complete)
PDF	Portable Document Format
SCORE	(not an abbreviation—a music score publication program)
SDG	Solo Dei Gloria
UK	United Kingdom

UNIX	(not an abbreviation—the name of a class of open source computer operating systems)
WTC	Well-Tempered Clavier (WTC1: Book 1 BWV 846–869, WTC2: Book 2 BWV 870–893)
XML	eXtensible Markup Language
ZIP	(not an abbreviation—a format for compressing computer files)

List of Figures

Fig. 2.1	Communication system from Shannon. From Shannon (1948) with permission of Nokia Corporation and AT&T Archives	10
Fig. 2.2	Noise in music copying. Diagram by the author based on Shannon	10
Fig. 2.3	Claude Shannon juggling on his unicycle and Bach at the organ. Left: Reproduced with kind permission of the Shannon Family. Right: "Bach, Orgel Spielend" by Edouard Hamman. From Hottinger, Christlieb Gotthold, Die Welt in Bildern (Orbis pictus), Berlin 1881, page 148. Scan by author	12
Fig. 2.4	Simplified database with hash table. Diagram by the author	14
Fig. 3.1	Musical alphabet. From Tatlow (1991). Reproduced with permission of the Licensor through PLSclear	21
Fig. 3.2	J.S. Bach as Rubin's Vase. Image by the author from "Joh. Sebastian Bach u. seine erste Gattin". From: Sammlung Fritz Donebauer, Prag: Briefe, Musik-Manuscripte, Portraits zur Geschichte der Musik und des Theaters; Versteigerung vom 6. bis 8. April 1908. Auktionskatalog. Berlin: J. A. Stargardt, 1908, S. 3. With kind permission of the Bachhaus Eisenach	28
Fig. 3.3	Hay's relation of notes and colours. From Hay (1838) Public domain archive.org. Digital image courtesy of Getty's Open Content Program	33
Fig. 3.4	Goldberg 30 + 2 by Benjamin Samuel. Goldberg Variation 30 + 12 (2010) © Benjamin Samuel Koren	33
Fig. 3.5	Licht's cryptographic coding table. From Licht (2009) with the kind permission of Deutsche Hochschulwerbung und-vertriebs GmbH	35
Fig. 4.1	Histogram of words per G-Value—modern German dictionary (Numeric Alphabet)	41

Fig. 4.2	Histogram of words per G-Value—modern German dictionary (Milesian Alphabet)	43
Fig. 4.3	Histogram of words per G-Value—modern German dictionary (Trigonal Alphabet)	43
Fig. 4.4	Comparative histograms of words per G-Value—Modern German dictionary	44
Fig. 4.5	Histogram of words per G-Value—German Luther Bible	45
Fig. 4.6	Histogram of words per G-Value—Cantata texts	46
Fig. 4.7	Histogram of combined historic sources—Latin natural	46
Fig. 4.8	Histogram of combined historic sources—Latin Milesian	47
Fig. 4.9	Histogram of combined historic sources—Trigonal	47
Fig. 4.10	Comparative histogram of combined historic sources	48
Fig. 4.11	Loss of information	49
Fig. 5.1	Partitions and multiple words	53
Fig. 6.1	Quantz (1752) Section XII §11V. With kind permission of Bärenreiter-Verlag Karl Vötterle GmbH & Co. KG.	58
Fig. 6.2	MIDI example—Extract from start of BWV 1001	59
Fig. 6.3	Processed MIDI data for first bar of BWV 1001	60
Fig. 6.4	MIDI data excel with notes in bar and G-values	61
Fig. 6.5	Score of first bar of BWV 1001. Staatsbibliothek zu Berlin—Preußischer Kulturbesitz. Mus.ms. Bach P 268. Public Domain 1.0. http://resolver.staatsbibliothek-ber lin.de/SBB0001DAD600000000	61
Fig. 6.6	Distribution of G-Values by bar for BWV 1001–1006 without repeats	66
Fig. 6.7	Distribution of no. of notes in bar for BWV 1001–1006 without repeats	66
Fig. 6.8	Example Encoding Procedure	71
Fig. 7.1	NATO study of probability statements. From (Internet33) (public domain)	75
Fig. 7.2	Shapes of normal and uniform distributions	76
Fig. 7.3	From mattheson phthongologia systematica p. 15. From Niedersächsische Staats- und Universitätsbibliothek Göttingen (Göttingen State and University Library) ID PPN684732122 with permission	78
Fig. 7.4	Effect of increasing sample size	81
Fig. 7.5	Example Bayesian network. Screenshots from the AgenaRisk program with permission—see Fenton (2013).	84
Fig. 7.6	Example Bayesian network—negative scenario	85
Fig. 8.1	Pythagoras and the Blacksmiths. From Quantz (1983, 2018) with kind permission of Bärenreiter-Verlag Karl Vötterle GmbH & Co. KG	88

Fig. 8.2	Leonardo da Vinci's interpretation of Vitruvius. ©G.A.VE—Archivio fotografico foto: Matteo De Fina, 2019 *"su concessione del MIBACT—Gallerie dell'Accademia di Venezia*	89
Fig. 8.3	Vitruvian man. From Rowland, Ingrid D. and Howe, Thomas Noble. Vitruvius Ten Books on Architecture. © Cambridge University Press, 1999. Reproduced with permission of the Licensor through PLSclear. Scanned by the author	90
Fig. 8.4	Wolff Anfangsgründe Vol. 1 p. 61—Equivalence of Proportions. Digitalisat des Universitäts- und Landesbibliothek Sachsen-Anhalt in Halle (Saale); ID VD18 90,183,886. http://digitale.bibliothek.uni-halle.de/urn/urn:nbn:de:gbv:3:1-433575 Creative Commons 3.0 with permission	91
Fig. 8.5	Some 1:1 proportions in Well Tempered Clavier Book 1	92
Fig. 8.6	Layers of proportion in numbers of bars	92
Fig. 8.7	Layers of 1:1 proportion	100
Fig. 8.8	Simple example of layer structure	101
Fig. 8.9	Output of solution search	106
Fig. 8.10	Diagram for example of 2 layers	107
Fig. 8.11	Output of solution search 1:1 with 2 layers	108
Fig. 8.12	Output of pattern matching	110
Fig. 8.13	Example colour coding of Mass in B minor	111
Fig. 8.14	Main output of Monte Carlo simulation	113
Fig. 8.15	Results histogram output of Monte Carlo simulation	114
Fig. 8.16	Lengths histogram output of Monte Carlo simulation	115
Fig. 9.1	Proportions in WTC1. From Tatlow (2015) reproduced with permission of the Licensor through PLSclear	118
Fig. 9.2	Infographics of WTC1 1:1 proportions within all combinations. Dartboard image by Vectorportal.com with annotations by author	121
Fig. 9.3	Second Layer in WTC1 from Tatlow (2015). Reproduced with permission of the Licensor through PLSclear	122
Fig. 9.4	Second layer in WTC1 1:1 as program output	122
Fig. 9.5	The first 1:1 two-layer solution with parallel number of pieces	123
Fig. 9.6	Distribution of individual lengths	124
Fig. 9.7	Monte Carlo simulation of WTC1 1:1	124
Fig. 9.8	Distribution of 1:2 Solutions for 24 Pieces	126
Fig. 9.9	WTC1 Monte Carlo simulations of 2 Layers	127
Fig. 9.10	Simple Bayesian network for 1:1 proportion in WTC1	128
Fig. 9.11	Node probability table for "1:1 in WTC1"	129
Fig. 9.12	Entering an observation in the Bayesian Network	130

Fig. 9.13	Bayesian network with result of observation of 1:1 proportion in WTC1	131
Fig. 9.14	Bayesian network testing the case of not finding 1:1 in WTC1	131
Fig. 9.15	Simple Bayesian network for WTC1 with sceptical prior	132
Fig. 9.16	Simple Bayesian network for WTC1 sceptical with observation	132
Fig. 9.17	Monte Carlo simulations of Ariadne Musica	150
Fig. 10.1	1:1 Solutions and probability against no. of pieces	164
Fig. 10.2	1:2 Solutions and probability against no. of pieces	165
Fig. 10.3	Convergence of probabilities for layers	166
Fig. 10.4	Lengths Histogram of Sei Soli 1:2	171
Fig. 10.5	Real versus Simulations 1:1 (Part 1)	175
Fig. 10.6	Real versus Simulations 1:1 (Part 2)	176
Fig. 10.7	Real versus Simulations 1:2 (Part 1)	177
Fig. 10.8	Real versus Simulations 1:2 (Part 2)	178
Fig. 10.9	François Blondel, column base and musical proportions. From Cohen, Matthew. Editorial Introduction: Two Kinds of Proportion in (Internet31). Creative Commons CC BY	180
Fig. 10.10	Example structure in Mass in B minor. from Smend (1950) with kind permission of Bärenreiter-Verlag Karl Vötterle GmbH & Co. KG	181
Fig. 10.11	Lengths of Bach's pieces and movements	191
Fig. 10.12	1:1 Proportions in Inventions and Sinfonias from Tatlow (2015). Reproduced with permission of the Licensor through PLSclear	193
Fig. 10.13	Tatlow's Multiple 1:1 Proportions of Inventions and Sinfonias as Output by Program	194
Fig. 11.1	Dieben's magic square for WTC1 (From Kramer (2000) with kind permission from the author)	206
Fig. 11.2	Part of WTC1 differences matrix	207
Fig. 11.3	Distribution of WTC1 length differences	208
Fig. 11.4	Two-layer 1:1 proportion on columns derived from Dieben	208
Fig. 11.5	Two-layer 1:1 proportion on rows derived from Dieben	210
Fig. 11.6	Two-layer 1:1 proportion on diagonals derived from Dieben	211
Fig. 11.7	Two layers of 1:2 proportion derived from Deiben rectangle	211
Fig. 13.1	Statues of Gottfried Wilhelm Leibniz and Johann Sebastian Bach in Leipzig. Photographs © by the author	218
Fig. 13.2	Mattheson's footnote on Bach and Mizler. From Mattheson (1740) https://archive.org/details/grundlage einereh00matt	219
Fig. 13.3	Standard model particles. Image by Dominguez, Daniel, © CERN, used with permission	221

Fig. 13.4	Spiral of perfect fifths	223
Fig. 13.5	Orthotonophonium in the Grassi Museum of musical instruments, University of Leipzig. Photograph © by the author with permission of the museum. Wiki Commons CC-BY-SA	224
Fig. 13.6	Structures of the Goldberg variations. Left: From Siegele (1997) with permission, modified and translated by the author. Right: By the author	225
Fig. 13.7	Magic rectangle from Mäser (2000). © Peter Lang used with permission	225
Fig. 13.8	Lull's concentric disc calculator. Lull (1274) Ars Magna https://archive.org/details/bub_gb_rG_yINh8V1gC/page/n25/mode/2up Public Domain Mark 1.0	226
Fig. 13.9	Harsdörffer's "Denckring" phrase calculator and circular slide rule. Schwenter (1692) Teil 2. http://digital.slub-dresden.de/id275480860 (Public Domain Mark 1.0)	227
Fig. 13.10	Leibniz' mechanical calculator reproduction. © Arithmeum, Rheinische Friedrich-Wilhelms-Universität Bonn. Reproduced with permission	228
Fig. 13.11	Quantz (1752) extract from XVIII §7. With kind permission of Bärenreiter-Verlag Karl Vötterle GmbH & Co. KG	231
Fig. A.1	Monte Carlo simulations of Mass in B minor Lutheran version	237
Fig. A.2	Monte Carlo simulations of Mass in B minor movements	239
Fig. A.3	Monte Carlo simulations of Brandenburg Concertos movements	241
Fig. A.4	Monte Carlo simulations of Brandenburg Concertos with all repeats	243
Fig. A.5	Monte Carlo simulations of Clavierübung II	248
Fig. A.6	Monte Carlo simulations of Clavierübung III	249
Fig. A.7	Monte Carlo simulations of French Suites movements	253
Fig. A.8	Monte Carlo simulations of Goldberg Variations	255
Fig. A.9	Monte Carlo simulations of Great 15 Organ Preludes	257
Fig. A.10	Monte Carlo simulations of Inventions and Sinfonias	258
Fig. A.11	Monte Carlo simulations of Musical Offering	260
Fig. A.12	Sei Soli movements 1:1—lengths histogram for 1000 samples	266
Fig. A.13	Sei Soli movements 1:1—results histogram for 1000 samples	267
Fig. A.14	Sei Soli movements 1:1—lengths histogram for 5000 samples	267
Fig. A.15	Sei Soli movements 1:1—results histogram for 5000 samples	268
Fig. A.16	Monte Carlo simulation of Sei Soli movements 1:2	271

Fig. A.17	Monte Carlo simulations of the concerto transcriptions	273
Fig. A.18	Monte Carlo simulations of Trio Sonatas movements	279
Fig. A.19	Monte Carlo simulations of Violin Sonatas movements 1:1	281

List of Tables

Table 2.1	Derivation of Gödel number	15
Table 3.1	Major and minor keys	25
Table 3.2	Summary of possible codings	35
Table 4.1	Summary of dictionary statistics	48
Table 6.1	Ways of counting bars in the Solo Violin Works BWV 1001–1006	62
Table 6.2	Occurrences of "Bach" gematria in BWV 1001–1006	65
Table 6.3	Occurrences of "non-Bach" gematria in BWV 1001–1006	65
Table 6.4	"Bach" gematria by numbers of notes in BWV 1001–1006	66
Table 6.5	Occurrences of "Maria Barbara Bach" = 95 in BWV 1001–1006	67
Table 6.6	Occurrences of "Maria Barbara" = 81 in BWV 1001–1006	68
Table 6.7	Occurrences of "Johann Sebastian Bach" = 158 in BWV 1001–1006	68
Table 6.8	Occurrences of "Mozart" = 87 in BWV 1001–1006	69
Table 6.9	Occurrences of "Beethoven" = 91 in BWV 1001–1006	69
Table 6.10	Occurrences of "Shepherd" = 80 in BWV 1001–1006	70
Table 8.1	Factors and proportions of 2088	94
Table 8.2	A simple set of pieces with numbers of bars	97
Table 8.3	Example solution pairs for 1:2 proportion	97
Table 8.4	Example binary signatures	99
Table 8.5	Examples of interesting patterns	100
Table 8.6	Demonstration with 6 simple lengths	105
Table 9.1	First 5 examples of the 14,191 different 1:1 proportions of WTC1	119
Table 9.2	Proportions of bars for WTC1	121
Table 9.3	Proportions of combinations in WTC1 if intentional	134
Table 9.4	Pattern matching input file for WTC1	137

Table 9.5	Combinations of WTC1 1:1 with same sequence in both halves	138
Table 9.6	Combinations of WTC1 1:1 with mirror image	139
Table 9.7	Combinations of WTC1 1:2 with mirror image	140
Table 9.8	Summary of solutions and patterns for WTC1	140
Table 9.9	Proportion of WTC1 preludes and fugues separately	141
Table 9.10	Proportions in WTC1 fugues	142
Table 9.11	Proportions in WTC1 preludes	142
Table 9.12	Kramer's and the symmetrical 1:2 patterns in lengths and pieces for WTC1 fugues	142
Table 9.13	First 8 of 5128 1:2 patterns in lengths and pieces for WTC1 preludes	144
Table 9.14	Symmetrical patterns in lengths and pieces with 1:2 in WTC1 preludes	146
Table 9.15	Example Left = Right pattern over 2 layers in WTC1 fugues	148
Table 9.16	Solutions and patterns in Ariadne Musica	149
Table 9.17	Summary of Solutions and Patterns in Ariadne Musica	151
Table 10.1	Short and full names of works and collections	154
Table 10.2	Number of single layer 1:1 solutions against number of pieces	156
Table 10.3	Number of strict two-layer 1:1 solutions against number of pieces	158
Table 10.4	Number of single-layer 1:2 solutions against number of pieces	160
Table 10.5	Number of two-layer 1:2 solutions against number of pieces	162
Table 10.6	Possible layers for works and collections	168
Table 10.7	Monte Carlo simulation of Sei Soli Works 1:2	169
Table 10.8	Monte Carlo frequencies for six pieces between 136 and 524 bars for 1:2	171
Table 10.9	Collections or works with distributions similar to normal	173
Table 10.10	Collections or works with too few samples	173
Table 10.11	Collections or works with few pieces	174
Table 10.12	Lengths of the pieces in Fig. 10.10	182
Table 10.13	WTC1 preludes & fugues durations proportions	184
Table 10.14	Goldberg Variations durations proportions	184
Table 10.15	Collections and works with no solutions	185
Table 10.16	Collections and works with no possible solutions	186
Table 10.17	Versions of the Art of Fugue	187
Table 10.18	Reverse engineering example art of Fugue	188
Table 10.19	Combining probabilities for multiple works	191
Table 10.20	Sei Soli and Violin Sonatas combined 1:2	195
Table 10.21	Monte Carlo simulation of Sei Soli and Violin Sonatas combined 1:2	195

Table 10.22	Summary of solution statistics	197
Table 11.1	Pieces in WTC1 with the same length	207
Table A.1	Summary of simple proportions in Mass in B minor Lutheran version	236
Table A.2	Proportions in Mass in B minor main sections	237
Table A.3	Simple proportions in Mass in B minor movements	238
Table A.4	Proportions in Brandenburg Concertos	240
Table A.5	Proportions in Brandenburg Concertos with all Repeats	242
Table A.6	Proportions in Canonic Variations	243
Table A.7	Monte Carlo simulations of Canonic Variations	244
Table A.8	Proportions in six Cello Suites	246
Table A.9	Partial results for A. M. Bach's copy of Cello Suites	247
Table A.10	Monte Carlo simulation of Clavierübung I works 1:1	247
Table A.11	Proportions in Clavierübung I movements	247
Table A.12	Proportions in Clavierübung II movements	247
Table A.13	Proportions in Clavierübung III	249
Table A.14	Monte Carlo simulation of English Suites works	250
Table A.15	Monte Carlo simulations of French Suites works	251
Table A.16	Proportions in the French Suites movements	251
Table A.17	Pattern in French Suites 1:1	252
Table A.18	Proportions in Goldberg Variations	254
Table A.19	Proportions in Great 15 Organ Preludes	256
Table A.20	Inventions and Sinfonias as separate works	257
Table A.21	Inventions and Sinfonias as one collection	257
Table A.22	Proportions in Musical Offering	259
Table A.23	Architectural proportion in Musical Offering 2:3	259
Table A.24	Proportions in Schübler Chorales	261
Table A.25	Monte Carlo simulations of Schübler Chorales without Repeats or Da Capos	261
Table A.26	Monte Carlo simulations of Schübler Chorales plus Da Capos	262
Table A.27	Monte Carlo simulations of Schübler Chorales including Da Capos	263
Table A.28	Monte Carlo simulation of Sei Soli 1:2	264
Table A.29	Monte Carlo simulation of Sei Soli 1:1	264
Table A.30	Proportions in BWV1001, BWV 1002 and BWV 1006	265
Table A.31	Sei solo movements 1:1 two-layer example	269
Table A.32	Sei Soli example of double parallel 1:1 in two layers of bars and pieces	270
Table A.33	Proportions in Sei Soli	272
Table A.34	Proportions in transcribed concertos	273
Table A.35	Partial result for Vivaldi's L'Estro Armonico	274
Table A.36	Left=Right symmetries in Trio Sonatas without repeats	276
Table A.37	First solutions of Trio Sonatas without repeats	277
Table A.38	Proportions in Trio Sonatas movements without repeats	279

Table A.39	Proportions in Trio Sonatas movements with repeats	279
Table A.40	Proportions in Violin Sonatas works	280
Table A.41	Proportions in Violin Sonatas movements	280
Table A.42	Proportions in Violin Sonatas movements with repeats	280
Table B.1	Pattern output indicators	297
Table C.1	Brief history of statistics	319
Table D.1	G-values of letters	323
Table D.2	G-values of notes	324
Table E.1	Musical intervals as proportions	325

Chapter 1
Introduction

1.1 The Science of Musicology

As indicated in the Preface, I have a latent interest in the ways various authors have found numerical aspects in music, in particular in that of Johann Sebastian Bach. Being rather sceptical about these, some of which border on mysticism, I wanted to apply some scientific thinking to the problem and try to find ways to separate the wheat from the chaff, the mystical and the speculative from the factual, and to show how serious research in this area can be aided by informatics and statistics.

The objective here is to show how these techniques can be used to explore an artistic field rather than to come up with any specific results; any results need to be interpreted in the light of their historical and musicological context, and that is beyond my qualification.

Musicology, like any other -ology, is a science, in this case applied to an art form. Mathematics and computation are basic tools of any science. Quantitative methods and statistics are a branch of mathematics, and computers only implement mathematical logic at multiple levels of abstraction.

I have occasionally been accused of spending too much time "on the computer". But I have never just been at the computer for its own sake, I use the computer to do something, such as develop a program, manipulate data, exchange messages, talk to someone, read the news, research in dictionaries, encyclopaedias and libraries, organise and view photographs, draw diagrams, watch videos, listen to music, write letters, essays and books, prepare presentations, perform banking tasks. Some people even play games on the computer!

The main difference between doing any of the above with a computer rather than without is that with a computer we can do any of these things much faster than with the original method. This makes it feasible to do things which would not have been feasible before. The computer is a unique advance in that it is so versatile (see list above), and that it is a tool for the mind rather than anything physical. These attributes set it apart from any other technological achievements.

© Springer Nature Switzerland AG 2021
A. Shepherd, *Let's Calculate Bach*, Quantitative Methods in the Humanities and Social Sciences, https://doi.org/10.1007/978-3-030-63769-9_1

To draw an analogy with a more basic tool, as the saying goes, if you have a hammer, everything looks like a nail. If you give a small child a hammer, it will play with it, experiment with it, hit things with it, throw it (thus reinventing an old Scottish sport!), and probably cause some damage and hurt itself. It is said that when Pythagoras came across some hammers being used by a blacksmith, he discovered the basic mathematics behind musical harmony.

A new technology always generates a lot of enthusiasm as well as scepticism and resistance at first, but sooner or later becomes part of the everyday toolkit. Computational musicology has been around since the 1960s, and mathematics has been applied to music since Pythagoras. But I have still heard it said that due to new computational and digital technologies we are at a new threshold in musicology.

Now that people are growing up with computers, we do not need to go looking for the musicological nails to which the computer hammer could be applied.

But it is very difficult to write a correct computer program of any complexity. One only needs to look at the errors corrected in the regular updates to any program or "app", and the news reports on the failure of computer systems on which we rely in everyday life, not to mention security breaches. Similarly, quantitative methods, statistics and probability are further tools at the scientist's disposal and are particularly prone to misinterpretation.

All scientists and engineers take compulsory courses in these subjects when they study, and so should musicologists. They will then know where computers, quantitative methods, statistics and probability are useful (and where they are not) and be able to do things that could not be done before while avoiding causing any damage and avoiding the pitfalls. Let us therefore marry these techniques with musicology, and carry musicology over the aforesaid threshold, if indeed it has not yet been crossed.

Of course, computers traditionally tend to be used in a brute force approach by trying thousands or millions of calculations in an exhaustive search. Results must be independently verified against common sense and traditional knowledge and insight.

There should not be a need for a separate branch of computational or digital musicology. Computer technology is just another tool available for the musicologist to use where appropriate to do what they have always done (or wanted to do) faster. One could follow other subjects and refer to "computer-aided musicology" if it needs specific mention.

This is part of the reason for choosing the title "Let's Calculate Bach". It also nicely links to Leibniz who, as we shall see, influenced many of the thinkers in Bach's circle.

Hopefully, this book will help to show some of the possibilities and pitfalls in my chosen area of study.

1.2 Numerology and Bach

As described by Tatlow (1991) and summarised in Tatlow (2015), speculation about numerical references in Johann Sebastian Bach's music started in the first half of the twentieth century—e.g. Werker (1922), Dieben (1939), Smend (1950)—and has continued ever since.

Tatlow (1991) gives a thorough analysis and history of numeric alphabets, poetical paragrams and their links to music in general and Bach in particular. She calls for caution in the application of the natural order alphabet in the analysis of Bach's music. Tatlow (2015) goes on from there to present new sources to show that there is a plausible historical basis for the existence of proportional relationships in music by Lutheran composers in the 1700s, and specifically within Bach's collections and works.

This book will suggest some tools and techniques from information theory, statistics and computer programming to extend the scientific basis for investigating and evaluating numerological theories relating to the music of J. S. Bach, or indeed, of any other composer.

The intention is to show how computer-based tools can be used to save time for researchers in this field. It is not the primary intention to obtain new results, although many examples of possible applications are given.

In some way, this is the opposite of "historically informed theory" formulated by Ruth Tatlow in Tatlow (2013). Instead of using historical sources in the context of the language and philosophy of Bach's place and time, we shall make use of modern tools and recent methods. These must not be regarded as competing with traditional methods but will hopefully provide some additional insights. It will be seen that many of these techniques have their origin before or during Bach's time and even in his geographical area, but this is not to say that he knew or used them.

I am not a qualified historian or musicologist, so any statements about composers' intentions or thought processes that I make should be taken as questions or suggestions. It is up to the musicologists to apply these methods, if they find them useful.

Other applications of mathematics to music, such as geometry, topology, set theory and group theory, are not in the scope of this book.

1.3 About This Book

The approach is to briefly present the theories, hopefully in a form suitable for the non-scientist, show how they can be applied in our context and then summarise the salient points for those wishing to use the methods. In the end, I would hope to be able to use the famous quote from Arthur Conan Doyle's character Sherlock Holmes: "You know my methods, Watson, apply them", although of course the methods are not mine but are in general scientific and mathematical use. I have tried to include sufficient detail for others to understand and reproduce the results, as is good scientific practice.

To start off, a little information theory is introduced in Chap. 2.

Chapter 3 uses this to point out where numerical messages or references could or could not be coded in music and gives some criteria for deciding which methods are feasible.

Starting with Smend (1950), the most common application of numbers to Bach has been in coding words with numeric alphabets. Chapter 4 examines the ambiguities behind coding words as numbers, and Chap. 5 extends this to phrases rather than single words.

Putting alternative messages in the notes of a musical composition was presented by (Thoene 2016) in a beautifully laid out book. Chapter 6 shows how a computer can be used to analyse the music score to find those and other messages in the score and examines some statistics.

Chapter 7 introduces some more statistical methods which we use in the subsequent chapters.

Chapters 8, 9 and 10 look at the proportions in the architecture of Bach's collections and works—proportional parallelism as introduced by Tatlow (2007, 2015)—and apply some mathematics to this. As this is the most promising area for research, I devote more space to this than to the above and use data from Tatlow (2015) to show what can be achieved with informatics in the humanities, particularly musicology. These more than other chapters make extensive references to Tatlow (2015), and I have tried to compromise between making this book readable independently, not duplicating Tatlow's work and ensuring that due credit is given to her research. There are certainly more places where references are not explicit.

Chapter 8 introduces some terminology and a computer program to explore various aspects.

Chapter 9 demonstrates how the various techniques can be applied using the example of J. S. Bach's Well-Tempered Clavier Book 1. Further data for other collections is placed in Appendix A and summarised with conclusions in Chap. 10. Section 10.15 is a brief look at magic squares or rectangles that have been strangely neglected, probably because they are not easy to analyse.

Chapter 12 points out some psychological fallacies in the hope that this will prevent the enthusiast from jumping to the wrong conclusions on whether numerical references were actually put there by Bach himself.

Chapter 13 points out some historical and geographical connections, showing how much of the knowledge applied here may have been available to Bach, and briefly tracing the history of statistics and computers. It also returns to the natural sciences to consider some parallels with these, the value of artificial intelligence and some other technical considerations.

An overall summary is given in Chap. 14.

Appendix A collects further detailed examples for Chap. 9.

Appendix B includes the full user manual for the Proportional Parallel Explorer program.

Appendix C summarises historical events from Chap. 13.

Appendix D shows the number alphabets used.

Appendix E lists some proportions for musical intervals.

Appendix F gives some Excel functions used.

1.3 About This Book

There are many tables and statistical graphs in the book. They are included for those interested in the details and to facilitate the reproduction of the results by others, as is good scientific practice.

Conventions

Chapters and sections are numbered hierarchically for easier orientation and cross-referencing.

As I have tried to avoid too much duplication of other work, there are numerous references to other literature and the Internet. These are given by author's name or "Internet" in round brackets and are listed in the Literature and Internet Sources at the end.

The references in the text as well as cross references between sections should be clickable if an electronic version of the text is being used.

Footnotes are indicated with symbols rather than numbers to avoid confusion with the use of superscripts to denote mathematical powers such as 2^2 for two squared.

Definitions

A prerequisite for a rigorous scientific discussion is to have a clearly defined terminology and to use it consistently. This is not as easy as I expected, as some musical terms have developed over centuries and have been used differently at various times. The following words are used throughout this book as defined here—these may differ from other usage. Where available and applicable, standard definitions from Grove (2001) are used. Other terms not defined here are considered to be clear enough, e.g. bar (or measure).

Some more terms will be introduced in Chap. 8.

Composition	The product of the activity or process of creating music (based on Grove (2001))
	As used here, a composition can be a piece, movement, work or collection
Piece	A self-contained composition from the initial clef to the terminating bold double bar line
	A piece can stand alone or be part of a work or be included in a collection
	Note: Grove (2001) points out the difficulties of pinning down a definition for "piece": "Attempts to pin down a more precise definition of a 'piece of music' are beset by philosophicaland semantic problems, ..."
Section	A part of a composition which is differentiated from other sections for some reason, e.g. by virtue of having a different tempo or a different key
	Note: A piece can have sections
	Note: A section can itself contain sections; for example, see "movement" below

(continued)

(continued)

Movement	A term for a section, usually self-contained and separated by silence from other sections, within a larger musical work Grove (2001) Note: I take the word "usually" to apply to both "self-contained" and "separated by silence". The movements in the masses or passions, for example, often follow without a silence, are sometimes separated only by a change in time signature and sometimes even overlap Note: Grove (2001) does not define "section" Note: A movement can have sections as defined above Grove (2001) also says: "It is unusual to speak of movements of larger works such as symphonies or sonatas as 'pieces'...", but for our purpose a movement is a piece as defined above and will become important from Chap. 8onwards Movements, therefore, are the pieces that make up cantatas, masses, passions, concertos, suites, partitas, sonatas and individual works such as the Goldberg Variations, the Musical Offering, the Art of Fugue
Work	A composition consisting of one or more pieces or movements, e.g. Two-Part Invention no.1 BWV 772 (consisting of 1 piece), Keyboard Partita no. 1 BWV 825 (consisting of 6 movements), Brandenburg Concerto no. 3 BWV 1048 (consisting of 3 movements), Mass in B minor BWV 232(consisting of 27 movements) A work by J. S. Bach has a BWV number
Collection	A set of works published together, e.g. Inventions and Sinfonias BWV 772-801(consisting of 30 works), Well-Tempered Clavier Book 1 BWV 846-869 (consisting of 24 works, each of whichhas 2 pieces, a prelude and a fugue), Clavierübung I (the Keyboard Partitas) BWV 825-830 (consisting of 6 keyboard works, each having multiple movements), the Brandenburg ConcertosBWV 1046-1051(consisting of 6 orchestral works, each with multiple movements)
Phrase	"A term ... used for short musical units of various lengths" Grove (2001) Note: Length here is not necessarily a whole number of bars as defined below Also used in this book in its linguistic sense
Length	The number of bars in a composition Note: This is not the same as the duration
Duration	The time taken to actually perform a composition Note: This varies between performances

The following is specific to this book:

G-Value	The gematria number derived by adding the numerical equivalents of letters of the alphabet, e.g. the G-Value of "Bach" with the Latin natural order alphabet is 14: BACH = B + A + C + H = 2 + 1 + 3 + 8 = 14

Chapter 2
An Information Theory Approach

2.1 Information and Communication

Human beings have always communicated with each other to exchange information, initially through gestures and sounds. As humans began to spread out in the world and needed to coordinate their activities, the need for communication over larger distances arose. Beyond the range of the voice, techniques such as drums, smoke signals, semaphore, fires, reflected sunlight, letters and gunshots came into use, and this was the state of the art up until the nineteenth century. With the advent of electricity, telegraphic communication through wires became possible over long distances, first over land, and then across the oceans. The invention of radio increased the range even further and enabled communication outside the world, into space.[1]

As these communications became ever more important to various aspects of life, military, emergency services, commercial, air and space travel, etc., they needed to be more reliable, not only less prone to failure and outages, but also less prone to errors in transmission, and they needed to be faster. The scientists and engineers trying to achieve this therefore developed theories of communication, and since the "stuff" being communicated is information, this involves theories of information.

Information theory is broadly concerned with the storage, processing and communication of information in the form of data. Storage can be regarded as communication through time—it is "sent" at one time and "received" or read at a later time.

Communication theory is concerned with the speed with which information represented by a message can be transmitted through a medium or channel, the effects of noise and distortion in the transmission channel, how to encode and decode the message to make optimum use of the channel's capacity and how to detect and even correct errors in the communication caused by noise or interference. A message in

---

[1] As chance would have it, I seem to have a particular affinity with this part of history—the first transatlantic cable was laid in 1866 by the SS Great Eastern, a ship designed by the great engineer Isambard Kingdom Brunel. The first wireless transmissions were achieved by Guglielmo Marconi from 1894 and his first transatlantic transmission in 1901. I studied Electrical Engineering and Electronics at Brunel University with a student apprenticeship at the Marconi Co. Ltd.

© Springer Nature Switzerland AG 2021
A. Shepherd, *Let's Calculate Bach*, Quantitative Methods in the Humanities and Social Sciences, https://doi.org/10.1007/978-3-030-63769-9_2

this context can be typed, written or spoken words, images, films, computer data, etc. A channel is a link between two points or parties through which messages can travel. A channel can use various media, not necessarily electronic as we saw above.

The implementation of the respective theories in practice is information technology and communications technology. The practical implementation of scientific theories in the form of technology is called engineering. Because digital electronic technology is extensively used in both information processing and communications, the fields of information science, information technology, communications science and communications technology have become intermingled; one popular acronym is ICT—Information and Communications Technology.

A first step in developing a theory about something is to be able to measure it. For communication and information, this was pioneered by (Hartley 1928) in 1928 and developed further by (Shannon 1948) in 1948.

2.2 Measuring Information—The Bit

Since communication is concerned with the transmission of information in the form of messages, let us start by thinking of a message as a sequence of symbols from the repertoire or the set of possible symbols that are known to both the sender and the recipient. The obvious example is a message consisting of an English language sentence made up of symbols from the alphabet. The quantity of information contained in a part of a message depends on the amount of "surprise" or reduction in uncertainty achieved by that part of the message—how unlikely it is that that symbol appears at that point in the message. For example, if you are receiving a message about the weather that starts "Today the weather is col", a "d" as the next letter would be highly probable and would not come as a surprise and would therefore contain little information—even if it was distorted beyond recognition or missing you would be confident that your guess of a "d" was right. But if the sentence went on to say: "Today the weather is coloured" (maybe as a poetical reference to a rainbow), the letters after "col" contain a large amount of information. Hartley defined the unit of information as depending on the number of possible symbol sequences on a logarithmic scale. Shannon (1948), developed this further, using logarithms to base 2, which very conveniently links the scale to the binary number system (adding 1 to the base 2 logarithm of a number doubles the number) and he used the term bit for the unit of information after John Tukey. Shannon also introduced the word "entropy" for the quantity of information, choice and uncertainty in analogy to its use in thermodynamics where it refers to the degree of chaos or lack of order in a system.

In written language, the information content of a letter depends on its frequency of use, e.g. "e" is the most frequently used letter and contains the least information. It also depends on preceding letters, e.g. a "u" after a "q" contains hardly any information because it nearly always occurs. This can be applied to musical notes, both for the frequency of their occurrence and their context of preceding or accompanying notes.

Information theory and entropy have been applied to music, for example, in comparing the styles of different composers or different musical genres, analysing melodic or harmonic progressions, or comparing music generated by artificial intelligence to samples from the composer it is intended to imitate. But these are not the applications with which we are concerned here.

2.3 The Bit as Binary Digit

As we saw above, the smallest unit of information is the bit. This can represent two different values: "yes" or "no", "on" or "off", "true" or "false," "1" or "0".

To convey or store more information, several bits can be strung together. Two bits can have the values "00", "01", "10" or "11", i.e. four different values. Each additional bit doubles the number of possibilities. These are binary or base two numbers. We usually count in decimal or base ten, i.e. with the digit positions from the right representing ones, tens, hundreds, thousands, etc., i.e. the powers of the base ten. In binary, the positions represent the powers of the base two, i.e. ones, twos, fours, eights, etc., and each position is a binary digit or bit.

Digital computers are implemented using binary arithmetic as this is much easier to implement in electronics than decimal arithmetic. We usually use bytes as the basic units. A byte consists of eight bits and so can represent 256 different values, and as computers always count from zero internally, the values are 0–255.

2.4 Signal, Noise, Redundancy and Encoding

Most communication channels introduce some kind of interference or noise and thus distort, falsify, or even obliterate parts of the signal. This is shown in the classic diagram from Shannon (1948) in Fig. 2.1. Applying this to the example of the analogue telephone, the information source is the person speaking, and the transmitter is the telephone mouthpiece which converts the sound waves into analogous electrical signals. These are transmitted along wires and through exchanges, but other electrical noise and interference occurs on its way to the handset of the receiver, which converts the electrical signals back into sound waves for the ear of the destination, the listener. The reconstruction of a signal or message distorted by noise depends on redundancy, i.e. the message consisting of more symbols than would be strictly necessary to convey the information alone. Spoken or written language contains a large amount of redundancy and we can understand speech in the presence of background noise at a party or dropouts on a bad telephone line. We can read text if letters are misprinted or missing, or even if the lower part of the line is obscured.

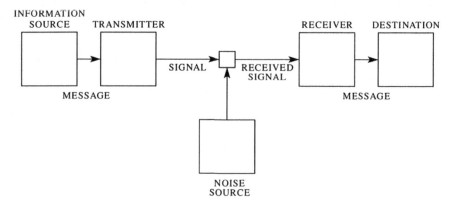

Fig. 2.1 Communication system from Shannon. From Shannon (1948) with permission of Nokia Corporation and AT&T Archives

Binary information stored or transmitted in digital computers and electronic communications systems usually contains some redundant bits which enable errors to be detected or even corrected.

The signal and noise aspects of information theory would apply to original musical (and other) manuscripts. The paper manuscript is the transmission channel and the written staves and notes are the message. Any smudges, faded ink, holes in the paper or missing pages, etc. are noise which interferes with the signal. As with written or spoken words, there is a certain amount of redundancy, which enables us to guess or even be certain of what a smudged or missing note should be. This also applies to errors made by the composer—a faulty transmission between the composer's brain and hand, or a copyist—a faulty transmission between the copyist's eye and hand—followed by centuries of wear and tear. This is shown as an adaptation of Shannon's diagram in Fig. 2.2.

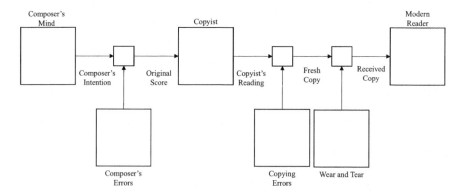

Fig. 2.2 Noise in music copying. Diagram by the author based on Shannon

To be transmitted through a channel or to be stored, information needs to be encoded, i.e. represented in some appropriate way. The most common information we use is written text in a language such as English or German with the corresponding alphabet as the symbol set. In electronic communications and storage, the letters, numbers, punctuation and other symbols are represented, i.e. encoded, in bytes. As we have seen, some redundancy is required to detect errors, so usually one or more extra (redundant) bits are added. Some codes can enable errors to be corrected. One common computer representation encodes A as 01000001, B as 01000010, C as 11000011 etc., which includes an extra parity bit (the first) to make the number of ones even—if a symbol is received with an odd number of ones, the recipient knows that it is wrong. This means that half of all the possible symbols are not part of the valid symbol set (This originated when the bits were transmitted serially—one after the other—so any interference or noise was likely to affect a single bit.).

Of course, the more redundancy that needs to be added to ensure correct transmission, the less net channel capacity is available for the actual message.

Coding can also involve encryption to ensure that the transmitted message can only be read by those for whom it was intended.

When considering possible numerological or other messages hidden in music, our transmission channel is the manuscript or printed score of the music including the words to be sung. The messages we are looking for are not the music itself, but other possible meanings. The encodings we are looking at are for the most part a numerical alphabet encoding A as 1, B as 2, etc., and there are different variants of these as we shall see, some of which have redundancy.

Having started to make the link between Shannon's information theory and the transmission of music, we can make a more specific connection to the composer in whom I am most interested, Johann Sebastian Bach. One of Claude Shannon's pastimes was riding a unicycle while juggling in the halls of the Bell Laboratories where he (and Hartley before him) worked. I do not know what Shannon's taste in music was, other than that he played the jazz clarinet Sonni & Goodman(2017)), but it seems to me that juggling while riding a unicycle is at about the same level of cognitive ability as playing a fugue with the hands on separate manuals and the feet on the pedals of an organ (Fig. 2.3).

2.5 Messages and Symbols

A message to transmit information must therefore be represented as a sequence of symbols. The available symbols are called the "symbol set". A textual message would normally consist of a word or phrase in a certain language (for J. S. Bach, this would be German or Latin, or even Italian or French). The message is composed of a sequence of symbols from the available symbol set, i.e. the letters of the alphabet. A message has a beginning and an end, and therefore a defined length. The processing, storage and communications equipment must be able to handle the specified maximum length of a message, including the overheads for redundancy.

Fig. 2.3 Claude Shannon juggling on his unicycle and Bach at the organ. Left: Reproduced with kind permission of the Shannon Family. Right: "Bach, Orgel Spielend" by Edouard Hamman. From Hottinger, Christlieb Gotthold, Die Welt in Bildern (Orbis pictus), Berlin 1881, page 148. Scan by author

The symbols can be encoded in various ways, for example, as numbers using numeric alphabets. To encode all the letters of the Latin alphabet A–Z, 26 symbols are required, or 24 if I & J and U & V are merged, as was common practice in Bach's time. As Tatlow has shown in Tatlow (1991), there are other alphabets (e.g. Hebrew or Greek) and many different ways of encoding alphabets. The simplest coding is to say $A = 1$, $B = 2$, etc., which is called the Latin natural order alphabet, and other examples are the Milesian and Trigonal (Tatlow, 1991). The codes are given in Appendix D.1, and later, we shall see what effect the choice of coding has.

Note: this is related to cryptography, which is intended to conceal the information from those who do not have the key, i.e. who do not know what encoding was used. Since we do not necessarily know what code a music composer may have used, code breaking techniques could be applied, but this is not pursued here. As Tatlow (1991) and (2015) show, we do know some of the coding devices used in the eighteenth century.

Another well-known but unrelated symbol set is the genetic code consisting of the four symbols A, T, G and C from which the messages contained in genes are built (see Hofstadter (1979) for some playful explanations).

2.6 Throughput and Protocols

The rate at which messages can be correctly transmitted through a channel depends on the capacity of the channel and the amount of redundancy required to compensate for the noise or interference. The throughput of a digital channel is usually given in bits per second or hopefully, if you are an Internet user, in megabits per second (millions of bits per second). This is of no concern here, as we are interested in static messages stored and carried through time in a musical composition. For storage, or transmission through an indefinite time, the capacity is measured in bits, or more usually bytes, megabytes, gigabytes, etc.

There are additional aspects needed in communications technology to ensure that a message is conveyed correctly, for example, indicators of when a message begins or ends, dividing messages into blocks of fixed length, using sequence numbers to keep parts of a message in the correct order and to ensure all parts are received, and replies for confirmation of correct reception or requests to repeat parts of the message if errors are detected. The conventions for this are called protocols and involve a little overhead which further reduces the effective capacity of the channel. For our purposes, we only need to consider the first point, recognising where a message starts and ends.

2.7 Gematria as Hash Coding

We have looked at the encoding of letters and words using numeric alphabets. Having encoded a word or phrase as a sequence of numbers, it is a small step to add the numbers up to obtain a single number that represents the word or phrase, for example, B A C H = 2 + 1 + 3 + 8 = 14. Such a double coding is called gematria. Since we will be referring to this often, I will call the resulting number the G-Value (for gematria) of the word or phrase. As Tatlow (1991) shows, using the number alphabet in poetical paragrams was a popular pastime, in particular using people's names.

In computer science, this double coding is known as "hash coding" or "hashing" and is used, for example, in indexing databases to accelerate searching; the hash code gives the position in the hash table where the pointer to the information can be found, thus avoiding a more time-consuming sequential search through the whole data table. A simplified example is shown in Fig. 2.4. The letters of "Beethoven" are added up, i.e. the hash function is addition, to give the G-Value 91; the ninety-first entry in the hash table shows that the data on Beethoven is at position 250 in the database.[2] For large, unordered databases, this is much faster than searching through the whole database directly.

---

[2]In real database applications, the binary representations of the characters are added up, not their positions in the alphabet.

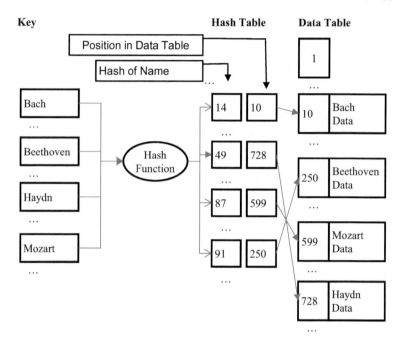

Fig. 2.4 Simplified database with hash table. Diagram by the author

Adding the numeric values of the letters like this means that several different words may add up to the same hash value. This requires some special handling in the above database application.

Hashing, or using gematria to represent a word, means that decoding back from the number to the original word can give an ambiguous result. We shall see the true extent of this ambiguity later in Chap. 4.

Putting it another way, hashing loses information and this lost information cannot be recovered. We can also say that this increases the entropy or amount of disorder. It is rather like whisking an egg—the original, more ordered structure of yoke and white is destroyed and this cannot be undone.

This is different to compression, where redundant information is omitted to save space, but the original message can be fully recovered on decompression; an example of this is the ZIP program which compresses computer files. There are also forms of compression that lose some of the information that is considered less important, e.g. some audio file formats such as MP3, or image file formats such as JPEG "lose" some aspects of the original message (music or video) that are considered to be imperceptible and achieve an even smaller file size.

Another form of hashing is used with telephone numbers. Each number on the telephone keypad is assigned to three letters, 2 is A B C, 3 is D E F, etc., familiar nowadays when typing text messages. This is used to make telephone numbers easier to remember, for example, 572224 could be given as JSBACH. This is only a one-way code, as there is too much ambiguity in the decoding.

2.8 An Unambiguous Coding

Hash codes or G-Values are ambiguous when we come to decode them, and the ambiguity can arise in different ways. The same letters in any order will give the same G-Value, i.e. all anagrams encode the same—B A C H or $2 + 1 + 3 + 8$ and C B H A or $3 + 2 + 8 + 1$ both encode to 14. Substituting letters, for example, such that one has the next higher G-Value and another has the next lower G-Value, will also give the same sum—B A C H codes the same as B A D G or $2 + 1 + 4 + 7$, and so on.

You may now ask how one can encode words with a numeric alphabet without ambiguity?

One answer to this was given by the mathematician Kurt Gödel in 1931 (although this is not why he did it). Bertrand Russell and Alfred North Whitehead tried to show that all mathematical truths could be derived from a set of axioms in their work Principia Mathematica. Gödel (1931) proved that this was not possible, and that for any such system, there would always be true statements that cannot be proved or disproved within that system. This is called Gödel's incompleteness theorem. To do this, he encoded the statements of Principia Mathematica into numbers, in much the same way as the G-Values encode words. The big difference is that Gödel introduced prime numbers[3] to fix the position of each letter in the word and used multiplication instead of addition. The resulting number is called the Gödel number of the expression. To take an example, we could encode the letters BACH using the numeric alphabet as follows—the steps are illustrated using Excel in Table 2.1:

(1) Take the numeric alphabet value for each letter,
(2) Assign a sequence of prime numbers to the letter positions,
(3) Raise each prime number to the power of the corresponding numeric alphabet value,
(4) Multiply all the results together[4] $2^2 \times 3^1 \times 5^3 \times 7^8$.

Table 2.1 Derivation of Gödel number

	Word	B	A	C	H
(1)	Numeric N	2	1	3	8
(2)	Primes P	2	3	5	7
(3)	P^N	4	3	125	5,764,801
(4)	$\Pi(P^N)$			8,647,201,500	

---

[3] A prime number is a whole number that cannot be divided by any other whole numbers (except for itself and 1), e.g. 2, 3, 5, 7, 11, 13, 17, etc.

[4] You may have seen the Greek letter sigma Σ used to signify the sum of a set of numbers. The Greek letter pi Π is used to denote the product of a set of numbers, i.e. multiplying them together rather than adding them up.

This obviously gives rise to very large numbers indeed and it will soon exceed the capabilities of Excel.

To decode the number, we simply need to find the prime factors of the coded number and count the occurrences of each prime number. It is a mathematical truth that every number has a single set of prime factors, i.e. prime numbers that give that number when multiplied together.

The prime factors of 8,647,201,500 are $2 \times 2 \times 3 \times 5 \times 5 \times 5 \times 7 \times 7 \times 7 \times 7 \times 7 \times 7 \times 7 \times 7$, i.e. 2 twos, 1 three, 3 fives and 8 sevens, giving us back the numeric alphabet values 2, 1, 3, 8 for BACH.

It is the use of prime numbers and the unique set of prime factors that makes the unambiguous decoding possible. Any other scheme, such as using powers of ten to fix the positions of the letters, would not work.

In fact, Gödel encoded the symbols using odd numbers so that he could differentiate whether a number represented a single symbol or a word. He also used the prime numbers above 13, i.e. from 17 onwards, as he used the first seven (including 1) for operators such as "=".

If you are interested in Bach and mathematics, art, genetics, etc., you might have read Hofstadter (1979), which combines Bach, Gödel, Escher and many other subjects, but does not treat this particular topic.

2.9 Codings and References

A numeric code does not have to represent the gematria or G-Value of a word but could refer to any significant number. I differentiate between:

- **Direct Codings**, where a word or phrase is represented in some coded way, e.g. "BACH" represented by "2" "1" "3" "8" or the sum of these numbers "14" which I call the G-Value.
- **Indirect Codings**, where a number is arrived at by some other means and is then interpreted (decoded) as a G-Value, for example, 14 entries of a theme decoding to "BACH".
- **References**, where a number occurring in a certain context, such as a piece of music, refers to something else or has a certain significance such as 3 for the Holy Trinity, 10 for the commandments and 11 or 12 for the apostles, or other numbers such as the year of an event. For example, Werckmeister writing in 1707 has a chapter "on the secret meaning of numbers" giving meanings to the numbers whose ratios give pure harmonies, 1, 2, 3, 4, 5, 6 and 8.

Tatlow (1991) gives more detailed classifications from the correspondence of Friedrich Smend including representative numbers, chronological numbers, symbolic numbers and numbers as quotations.

We do not need more detail here than to differentiate between codings and references.

Chapter 3
Some Possible Codings in Music

3.1 Preamble

Now let us consider how information other than the music itself could be encoded in a musical score, how much information could be thus encoded, and what information this could be.

To successfully transmit a message the following criteria must be fulfilled:

- The encoding of the symbols must be known (Sect. 2.4).
- There must be a sufficiently large symbol set to encode a message, at least 24 symbols to encode the eighteenth century alphabet (Sect. 2.5).
- The channel must have sufficient capacity to carry a message of useful length (Sect. 2.6), e.g. the message "BACH" has a length of 4 symbols. References only need one symbol—see Sect. 2.9.
- If the message is embedded in other irrelevant symbols, it must be possible to recognise the start and end of the message (Sect. 2.6), and indeed to recognise that there is a message at all.

Smend—see Tatlow (1991, p. 26)—lists some possibilities for Bach's use of numbers, e.g.:

Number of[1] movements, bars, parts
Number of statements of a theme
Number of notes in a theme
Number of notes in the bass
Number of notes in the instrumental voices of chorale passages.

---

[1] In computer science circles, it is usual to abbreviate "the number of" with "#", but this is avoided here both for the sake of the non-computer scientist reader and to avoid confusion with the musical sign for "sharp".

Here, these are extended and considered in more detail in the following sections to determine how well they fulfil the criteria given above.

The results are summarised in Sect. 3.20.

3.2 Number of Bars

For direct codings, the number of bars in a piece could represent an arbitrarily large symbol set, since a musical composition can have virtually any number of bars. Short pieces or movements of a few bars are less frequent, but do occur, e.g. as a puzzle canon which is not worked out, or some short recitatives in choral works which can be as short as 2 bars or the one-bar middle movement of the third Brandenburg Concerto BWV 1048. Two of the longest in Bach's works are the first movement of the fifth cello suite BWV 1011 with 446 bars (including repeats) and the fugue from the second sonata for violin solo BWV 1005 with 354 bars.

Each piece or movement has a certain number of bars and so can only encode one symbol, and larger works could encode a usable message length, e.g. the St. Matthew Passion BWV 244 with 68 or 78 movements, depending on which edition is used. The start and end of a message would only be easily recognisable if the message used the whole work or collection. I have not seen any claims that messages have been encoded in this way, and the obvious encoding of BACH in a work of 2, 1 3 and 8 bars is not viable.

Some examples of numbers of bars being references to significant numbers are given in Prautzsch (2000) and Hirsch (1986). Kramer (2000) finds extensive meanings of bar numbers and their sums in WTC1 (the Well Tempered Clavier Book 1) based on Dieben (1940).

Rumsey (1997) points to an anonymous source which arranges the numbers of bars in WTC1 in magic squares—more on this in Sect. 10.15 (I later identified the source to be (Dieben 1954) from Mäser (2000) and Kramer (2000).).

There is no way of knowing if any of these occurrences are intentional. If this was used extensively, one would expect to see bar numbers written in the manuscripts as it is easy to miscount, but this does not appear to be very widespread: Tatlow observes in Tatlow (2015, p. 120): "Knowing the number of bars in a movement or a section helped create the best possible layout of the copied page, and was also a way of checking that the copy was identical in length to the original. Copyists frequently kept track of the number of bars at the end of a movement. Bach's annotations occasionally survive in his scores—some were caused by copying, but others are not so simple to explain".

The numbers of bars in a composition can also be used to create proportions between the lengths of its components rather than coding any information in the number of bars—we shall return to this in Chaps. 8 and 9. This has been covered extensively in Tatlow (2015).

I did consider it possible that there would be some conventions on the lengths of compositions (e.g. 12 bar blues in jazz), especially for dance movements, and Tatlow

3.2 Number of Bars   21

(2015, p. 111) does point some out, including the fact they called for proportions, the significance of which we shall see later. Of course, a formal dance would have to have had a fixed number of bars so that the choreographed routine would end with the music, but the dance movements in suites and partitas were probably not intended to be danced to in Bach's time.

By extension, the number of beats could be used. These are simply a multiple of the number of bars depending on the time signature, e.g. four beats to the bar for 4/4 time, three beats to the bar for 3/4 or waltz time, etc.

3.3 Notes

The notes themselves can be used for direct coding. There are only 7 note names A–G or 8 in German A–H.[2] The obvious use of this is the notes B A C H, used both by Bach himself and later composers in homage to him. Dmitri Shostakovich is known to have used notes to encode the first letters of his name DSCH as a theme, and Brahms and Schumann also used this technique.

This can be taken further by going beyond the octave, e.g. with each note on the stave representing a letter or other symbol (see Fig. 3.1, an example reproduced from Tatlow (1991, p. 103) with Porta 1583, Schwenter 1620, Kircher 1650 and Schott 1665. There is another in Tatlow (2015, p. 63). The three staves of an organ work could encode a set of about 35 symbols by the position on the staves (more by adding more octaves).

A keyboard work with chords would be confusing as the sequence would not be clear, but a work for choir and/or orchestra has more potential as there are more than two or three staves mostly with only one voice on each.

The length of the message that can be encoded is limited only by the number of notes in the work.

The start and end of the message cannot easily be recognised, unless the message starts with the first note. The phrase B A C H is quite easy to recognise once one

Porta	b	a	c	d	e	f	g	h	i	k	l	m	n	o	y	z	r	s	t	u	w	x	q p
Schwenter	b	a	c	d	e	f	g	h	i	k	l	m	n	o	y	z	r	s	t	u	w	x	q p
Kircher	a	b	c	d	e	f	g	h	i	l	m	n	o	p	q	r	s	t	u	x	y	z	
Schott	a	b	c	d	e	f	g	h	i	k	l	m	n	o	p	q	r	s	t	v	w	x	y z

Fig. 3.1 Musical alphabet. From Tatlow (1991). Reproduced with permission of the Licensor through PLSclear

---

[2] In German notation, B is B♭ and H is B♮. Tatlow (2015) points out that in Bach's time both B and H were interchangeably used for B♮.

is aware of it (In my opinion, it is not a particularly pleasant tune, especially in a baroque context.).

The problem with this type of encoding is that the resulting music would probably not be pleasant, unless the coding was chosen after the piece was written and without coding the letters in alphabetical sequence of notes in the scale. However, this would make decoding very difficult if the coding scheme was not provided, making it more like encryption. We will revisit this in Sect. 3.19.

Kramer (2000) notes that copying errors would destroy any message and so this would be less likely to be used and this harks back to noise and redundancy which we treated in Sect. 2.4.

Indirect coding could be used, e.g. by summing the G-Values of the notes, which will come up in Sect. 3.8. This would get around the above problem of the quality of the tune, as the order of the notes would not be important, but still has the difficulty of denoting the start and end.

3.4 Intervals

An interval is the difference between two notes, e.g. C to E is a (major) third. Rumsey (1992) notes that the 14 intervals of the theme of the Passacaglia in C minor BWV 582 add up to 43, the G-Value of "CREDO". This is an example of indirect coding. We cannot say whether this was intentional or not.

For direct coding, taking the eight notes of the diatonic scale gives a symbol set of 8 values from the interval "first" (between a note and itself) to "eighth" or octave. Taking the interval in semitones would give 11 symbols. This can be extended beyond the octave, and intervals of "ninth", "tenth", etc. are also used. This could be taken to 24 by spanning over two octaves, similarly to Fig. 3.1.

The length of the message is only limited by the number of notes in the work.

There is no easy way of denoting the start and end of the message, unless as in Rumsey's example the message starts with the first note.

Incidentally, the Parsons code uses a simpler representation to show whether each note goes up or down or repeats the previous note without regard for the duration or the actual interval. This is used to identify melodies with a computer online, e.g. in (Internet7).

3.5 Note Lengths

The values of the notes could be used, i.e. breve, semibreve, minim, crotchet, quaver, semiquaver, demisemiquaver, etc., giving 7 or more options for each note, i.e. a set of 7 symbols, not enough to encode the alphabet. Including dotted notes could give further differentiation.

3.5 Note Lengths

The length of the message is effectively unlimited. The potential use is restricted, as the total length of the notes in a bar must be correct for each voice. There is no way of showing the start and end of the message.

An example is shown in the frontispiece, where the lengths in quarter notes are shown by the number of abacus beads over each note.

3.6 Number of Notes

The number of notes in a musical work, section, bar or even phrase can be used to encode symbols, e.g. a number alphabet. Since this does not limit the pitch and length of the notes, it is relatively easy to incorporate the desired coding into a valid and pleasant musical composition. It is also no problem to encode a sequence of symbols, e.g. in consecutive bars, to incorporate a word or a multiple-word phrase. The symbol set is only limited by the number of notes that can fit in a unit such as a bar, and with polyphony this is more than enough to cover the alphabet. The length of the message is only limited by the length of the work.

An example is given in Tatlow (1991, p. 105) and Tatlow (2015, p. 57) in which the composer Johann Christoph Faber coded the letters of the name of the dedicatee of a composition with the number of notes played by the trumpet in each movement. An exhibit in the Bach Museum in Leipzig which I saw in 2016 showed "Alles mit Gott und nichts ohn' ihn" BWV 1127, an air by Bach congratulating Wilhelm Ernst, Duke of Sachsen-Weimar on his birthday and noted that the instrumental introduction contains 52 notes representing the 52 years of the duke. Another possible example is given in Gardiner (2013) p. 422 in the St. John Passion BWV 245 movement "Gebt mir meinen Jesum wieder" where 30 notes in the violin could represent Judas' 30 pieces of silver. The subject of the first fugue in the Well Tempered Clavier Book 1 BWV 846 has 14 notes (BACH). These are references rather than codings. From the explicit context of these findings, it is plausible that they were intentional.

Extensive examples of numbers of notes used as references are given in Hirsch (1986). Thoene (2016) gives examples of the number of notes in the violin Ciaccona from Partita no. 2 BWV 1004 representing G-Values. Chapter 6 will show how a computer can be used to explore the plausibility of the latter.

The concrete examples above show the start and end of the message by placing it in one voice (the trumpet) or as the introduction of the piece. The last example uses phrases without any obvious start or end criteria, which, as we shall see in Chap. 6, further detracts from its credibility.

3.7 Number of Pieces, Movements or Sections

The number of pieces, movements or sections in a work or collection varies depending on the nature of the composition, e.g. 1 for a fantasia, 3 for a concerto, about 6 for a suite or partita, 32 for the Goldberg Variations BWV 988, 24 or 48 for the Well Tempered Clavier BWV 846-869, 27 for the Mass in B minor BWV 232, 64 for the Christmas Oratorio BWV 248 or 68 for the St. Matthew Passion BWV 244. There is therefore an adequate symbol set, but only one symbol can be thus encoded in a single composition, a maximum message length of 1.

Smend (1966) (page III.18) and Gardiner (2013) (p. 295) point out that cantatas BWV 75 "Die Elenden sollen essen" and BWV 76 "Die Himmel erzählen die Ehre Gottes", Bach's debut cantatas in Leipzig, each contain 14 movements, which is an indirect coding of his own name. Gardiner (2013) (p. 375) shows that there are 10 choruses in the trial scene of the St. John Passion BWV 245 which could represent a connection between secular law and the Ten Commandments, which is a reference rather than a direct coding.

Tomita (1999) also points out some numerical references in the Inventions and Sinfonias and the Well Tempered Clavier:

> It would not be totally surprising that beneath what appears as a methodically assembled architecture there could be shared theological concepts as a fundamental framework of the work. There is a yet-to-be-proven theory, however bold and speculative, that the number symbolism can be considered present in both works. In the Well-Tempered Clavier, on the one hand, the number '24', as in the number of pieces therein, can be interpreted as 24 seats and elders around the throne (Revelation 4:4). The number '30' in the Inventions and Sinfonias, on the other hand, can be interpreted as the number of years Jesus spent with his parents to learn the basics of mankind. By referring to these two holy numbers that express these specific theological meaning, Bach may have implied that the two-fold pedagogical process had a dual purpose—namely the fundamental work and further advanced work for those who completed the fundamental work in order to find pleasure in study and for the glory of God. This theological allusion in a didactic situation conforms to the attitude of musicians who lived in the Lutheran tradition of music education.

A composition may have a number of sections, e.g. Prelude and Fugue in E♭ major BWV 552—both the prelude and the fugue have three sections said to represent the Father, Son and Holy Spirit of the Trinity. Schütz (1941) finds this too trivial and finds many other reasons why this work refers to the Trinity. However, other works have three sections, e.g. in Clavierübung I the Sinfonia to keyboard Partita no. 2 in C minor BWV 826, but there is no reason why this should be associated with the Trinity. It is therefore impossible to tell whether there is any intentionality here—even if there was, Bach would probably not have taken care to avoid having three sections when he did not intend to refer to the Trinity.

Since we are considering the structure of complete works or collections here, the start and end of the message is clear.

3.8 Sum of the G-Values of Notes

The names of the notes (independent of octave) can be taken as number alphabet values (A = 1 to H = 8, and C♭, in German Ces = C + E+S = 33 etc.). The symbol set is shown in Appendix D.2 and has 34 symbols—more than enough to represent the entire alphabet, even though some, such as His (B ♯), are hardly in normal use. With the Latin Natural and the Milesian alphabets, Cis (C ♯) and Ges (G ♭) both have the same G-Value. The resulting numbers are not sequential and have gaps, so a direct encoding cannot be used.

For indirect coding, the note values can be added to give a number which is then interpreted as the G-Value of the intended word. The length of a message is arbitrary. This can be used by adding the values of the notes in a bar or phrase to represent the letters or words of a message. See Thoene (2016) for examples, such as on p. 31 the notes in the last bar of the fugue of the Sonata for Solo Violin BWV 1011 add up to 158 (Johann Sebastian Bach). With this usage, any length of message can be encoded, but the start and end of the message cannot be determined except for hints if it matches bars or phrases, or perhaps in Thoene's example as a signature in the last bar.

This is a sort of hash coding, and although virtually any number of symbols can be thus encoded, the encodings are not unique, so decoding has ambiguity. The extent of this will be made clear in Chap. 4.

3.9 Key Signature

There are 35 different major and minor keys, represented by between 0 and 8 sharps or flats in the key signature (see Table 3.1—the left column shows the number of

Table 3.1 Major and minor keys

No. of ♯ or ♭	Flats		Sharps	
	Major	Minor	Major	Minor
0	C	a	(C	a)
1	F	d	G	e
2	B♭	g	D	b
3	E♭	c	A	f♯
4	A♭	f	E	c♯
5	D♭	b♭	B	g♯
6	G♭	e♭	F♯	d♯
7	C♭	a♭	C♯	a♯
8	F♭	d♭	G♯	e♯

sharps or flats in the key). C (no sharps or flats) is only counted once in major and minor. 8 flats is actually represented by six flats and one double flat ♭♭. 8 sharps uses 6 sharps and one double sharp 𝄪.

Individual sharps and flats on various notes cannot be used fully, as not all combinations are musically correct, e.g. a single sharp on G is not a valid key signature. There are 17 valid key signatures (0–8 sharps and 0–8 flats). Each key signature can represent a major or minor key giving 34 possibilities for the symbol set, but major or minor is not encoded in the key signature itself and must be read from the title or content of the music.

Since a piece usually only has one key signature, the possible message length is limited (There are exceptions where the key is changed within the piece, e.g. the Ciaccona from Partita no. 2 for solo violin BWV 1004.).

Tatlow (2015) shows that Bach used the keys B–A–C–H or just B–A–C, sometimes in permutation, as keys of works in collections or as keys of movements in works.

One obvious use of this would be for Bach to have used a key signature of 3 sharps or 3 flats as references in his Trinity cantatas, as 3 is representative of the Holy Trinity. The cantatas for trinity are [see Dürr (2010)] BWV 165 "O heiliges Geist- und Wasserbad" in G major (1715), BWV 194 "Höchsterwünschtes Freudenfest" in B♭ major (1723/1724) and BWV 176 "Es ist ein trotzig und verzagt Ding" in C minor (1725). Only the last has 3 flats. This could be further considered regarding the time when Bach might have started coding information in his music, i.e. whether he only learned of paragrams around 1725 before he composed BWV 176 and after he composed the previous two, however it is possible that Bach encountered these as early as 1713 [see Chap. 5 of Tatlow (1991)].

The Prelude and Fugue in E♭ major (3 flats) BWV 552 is interpreted as a reference to the Holy Trinity, and some commentators take this to be intentional (see Sect. 3.7).

Böß (2009) even counts the number of sharps over all the staves in the C♯ major preludes and fugues in WTC1 and WTC2 to arrive at a number (1148) which he interprets as $2 \times 14 \times 41$ (14 = BACH and 41 = JSBACH)—an indirect coding.

The keys have been given various emotional associations (*Affekte*), however these depend on the tuning used, are personal and often contradictory, and would not represent any specific message. In modern equal tempered tuning, there is no difference between the keys.

In orchestral works, Bach's choice of keys is more likely to have been influenced by the choice of instruments, e.g. horns and trumpets being more easily playable in F or D.

If the message spans a whole work or collection, the start and end are obvious. In Tatlow's example of 3 works in the collection of 6 solo violin works having keys of B A C, the position of the message is given by the proportions, although with an extraneous G minor. In the example of CÜ I and CÜ II the pieces with keys of B C A H are separated by other pieces and there is no other indication of the location of the message.

With the single key signatures there is nothing to indicate that this is a message at all.

3.10 Accidentals

There are 7 possible symbols:

- sharp ♯
- flat ♭
- double sharp 𝄪
- double flat ♭♭
- natural flat ♮♭
- natural sharp ♮♯
- and natural ♮

This is a small symbol set of 7 symbols, and it cannot be used freely within a valid composition, although their number (the length of the message) is not limited.

3.11 Occurrences of Words

This was considered by (Smend 1966) in combination with the number alphabet, e.g. "CREDO" = 43 appearing 43 times in the Mass in B minor. However, this usage is a reference, which could be taken to emphasise the word, rather than a coding.

The number of words in a choral composition or the number of repetitions of a word could be used similarly to the number of bars (Sect. 3.2) or the number of notes (Sect. 3.6), and the possible message length is limited in the same way only by the length of the composition. It is not difficult to phrase a sentence such that it incorporates the desired number of words, e.g. "This sentence has five words"; "This sentence now has six words".

The criterion of having a recognisable start and end of the message can be fulfilled by the entire message being contained in a movement of composition.

3.12 Rests

One could consider the rests or silences to be the background against which the musical notes are perceived. Similarly to an optical illusion where the picture and background can be interchanged, as in the well-known example in Fig. 3.2 known as "Rubin's Vase" which can be seen as two black faces in profile on a white background or a white vase on a black background[3], so too could one consider looking for a message in the rests rather than the music.

There are 8 rest symbols from Long, Breve, down to demisemiquaver, equivalent to the note values in the Notes Sect. 3.3. This gives a limited symbol set.

---

[3] As I was not able to find the copyright owner of the original Rubin's Vase in Rubin (1915), I made my own from the silhouette of J.S. Bach.

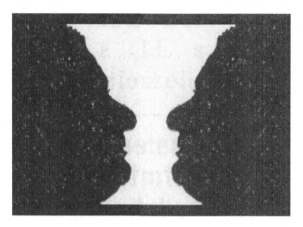

Fig. 3.2 J. S. Bach as Rubin's Vase. Image by the author from "Joh. Sebastian Bach u. seine erste Gattin". From: Sammlung Fritz Donebauer, Prag: Briefe, Musik-Manuscripte, Portraits zur Geschichte der Musik und des Theaters; Versteigerung vom 6. bis 8. April 1908. Auktionskatalog. Berlin: J. A. Stargardt, 1908, S. 3. With kind permission of the Bachhaus Eisenach

The message can be of arbitrary length, although some sequences of rests would not be correct or normal practice. As with the lengths of notes, the total length of rests and notes together in a bar must be correct for each voice, further restricting the possibilities. Here again we have difficulty in recognising the start and end of the message unless it starts with the composition.

3.13 Time Signature

The two numbers of the time signature cannot be freely used due to musical convention. The range of time signatures used in Bach's time was less than today, where mixed and irrational metres are used. There are rare examples of Bach changing the time signature within a piece (see Tatlow (2015, p. 25)) and Bach did use some less common time signatures; for example the following are those used in the Goldberg Variations (see Williams (2001)):

2/4, 3/4, 4/4,
3/8, 6/8, 9/8
12/8, 12/16, 18/16.

Other more or less common time signatures are 2/2, 3/2, 5/4, 6/4, 9/4, 12/4, 2/8.

About 16 possible time signatures, which normally only occur once in a piece or movement, do not provide a very large symbol set, and the length of the message is limited to 1 or very few symbols per piece or movement or up to about 68 for a work (see Sect. 3.7). The start and end of the message would be clear if the message spans the whole collection or work.

3.14 Figured Bass

In Bach's time improvisation was common and expected. The accompanying chords to an orchestral or choral piece to be played on a harpsichord, organ, lute or similar instrument, called the continuo, were not written out in full. The continuo score usually only contains the bass note and some numbers to indicate the harmony using the numerical intervals of the scale, e.g. 3 5 for the third and fifth denoting the major triad.

(Internet8) lists 39 different figures, which is a sufficiently large symbol set to code an alphabet, however, the sequence of symbols is restricted by the rules of harmony and harmonic progression. The length of the message is limited only by the length of the composition. The symbols could be either the different harmonies or the individual numbers in the figured bass (or their sums), the latter being more restricted. There is no obvious way of denoting the start and end of the message.

Tatlow (1991, p. 110) mentions this, and states that no example of its use for coding messages could be found.

3.15 Entries of a Theme

The number of entries of a theme, e.g. the subject of a fugue, could be used to encode a single number, e.g. Fugue 1 of the Well Tempered Clavier Book 1 BWV 846 has 24 entries (not all complete) which could be taken as a reference to the 24 preludes and fugues in the book. Marissen (2016) also points out the possible significance of the ten trumpet entries in BWV 77 "Du sollt Gott, deinen Herren, lieben".

More than one number could be encoded, e.g. 2 numbers for a double fugue.

The symbol set is only restricted by the possible number of entries of a theme in a piece and can easily accommodate enough symbols for the Latin natural order alphabet (Fugue 1 above would be "Z"). The possible length of the message is 1 per fugue (or 2 for a double fugue), e.g. 48 for the WTC books I and II together. The message would span the whole composition making the start and end clear.

3.16 Other Possibilities

3.16.1 Acrostics

Additional messages in the text of a cantata can be formed with an acrostic, i.e. the first letters of each line forming a word or phrase. An example is an aria in BWV 150 "Nach dir, Herr, verlanget mich", the first letters of which give the name of the presumed dedicatee Doctor Conrad Meckenbach.

This is only clear when some copying errors are corrected, (see Schulze (2017) – note that Dürr (2010), the last revision of which was in 1995, does not have the corrected text), and these errors are an example of redundancy alerting the musicologist to a probable error in transmission—see Sect. 2.4.

...

Doch bin und bleibe ich vergnügt,
Obgleich hier zeitlich toben
Creütz, Sturm und andre Proben,
Tod, Höll und was sich fügt.
Ob Unfall schlägt den treuen Knecht,
Recht ist und bleibet ewig Recht.
Cedern müssen von den Winden
Oft viel Ungemach empfinden
Niemals werden sie verkehrt.
Rat und Tat auf Got gestellet,
Achtet nicht, was widerbellet,
Denn sine Rot ganz anders lehrt.

...

Meine Tage in dem Leide
Endet Gott dennoch zur Freude:
Christen auf den Dornenwegen
Kühren Himmels Kraft und Segen.
Bleibet Gott mein treuer Schutz,
Achte ich nicht Menschentrutz;
Christus, der uns steht zur Seiten,
Hilft mir täglich sieghaft streiten.

Another example is the Ricercar from the Musical Offering BWV 1079, where the initial letters of the title "**R**egis **I**ussu **C**antio **E**t **R**eliqua **C**anonica **A**rte **R**esoluta" give the word "RICERCAR" (meaning "seek" as in "seek and ye shall find" as a hint to look for hidden messages).

The start and end of the message are clear, as it starts with the poem or title. The actual presence of the message must be noticed by the reader unless the relevant letters are emphasised in the typesetting or a hint is given as in "ricercar".

3.16.2 More Subtle Ways

A less direct way of conveying a message is treated in Marissen(1995) where the choices of instrumentation are seen as a commentary on social hierarchies. This is not a direct encoding of a text or a reference, but a much more subtle message.

3.17 Beyond Bach

3.17.1 BWV Numbers

As the Bach Werke Verzeichnis (Index of Bach's works) BWV numbers were only invented long after Bach's death, it would be very arcane to attribute any significance to these, although this might not deter some enthusiasts.

Each work can only encode a message of one symbol, and the symbol set currently consists of 1128 different values.

A concert programme consisting of a sequence or works could be used to encode a message of a few letters.

3.17.2 Frequencies

Musical notes have pitches or frequencies, e.g. $A^4 = 440$ Hz. Early means of measuring frequency were invented during Bach's time, e.g. the Savart wheel in 1681. The tuning fork was invented in 1711 by G.F. Händel's trumpeter John Shore and this had a frequency of $A = 432.5$ Hz, although it is not clear whether he actually measured this. See (Internet9) for an extensive treatment.

I have seen no indications that Bach knew of or concerned himself with these.

There is much diversity in the definition of concert pitch and in Bach's time there were different pitches for chamber and church music. There are also adjustments for different temperaments, which were tuned by ear hearing the beats per second of one note relative to another, such as an octave or a fifth. The frequencies of the notes of the scale were therefore by no means fixed. The frequencies are mostly not whole numbers. If the frequencies themselves are not used as numbers but allocated to symbols, using these would in effect, be equivalent to notes as in Sect. 3.3.

I know of no historical evidence that Bach was able to measure frequencies, but this has not deterred authors such as (Böß 2009) from finding gematric links using numeric alphabet interpretations of various word combinations that add up to 435 or 145 as the frequencies of notes to claim that Bach tuned to $a^1 = 435$ Hz and or $d^0 = 145$ Hz.

The large subject area of tuning and temperament is beyond our scope other than a further mention in Sect. 13.2.

3.17.3 Morse Code

(Internet10) mentions the use of Morse code in modern music. Morse code uses sequences of short and long tones (dots and dashes) to represent letters and numbers, and so could be transmitted in short and long notes in music. To be officially correct it would have to observe the timings, a dash being three times the length of a dot, gaps between symbols the same length as a dash and the gaps between words the length of seven dots (Internet11)

It was invented in 1837, so the comment in 3.17.1 applies here as well. The best-known example is probably the interpretation of the first four notes (three short and one long) of Beethoven's fifth symphony as Morse code for "V" for victory in world war II.

Incidentally, the Morse coding was chosen to make the best use of the available channel capacity by giving the most frequently used letters (in English) the shortest codings, e.g. the most frequently occurring letter "e" is one dot.

In conventional Morse code transmission, the message is preceded and followed by silence, so that the start and end are clear. Most occurrences in music start the message at the beginning of the composition.

3.17.4 Colours and Shapes

It is possible to draw parallels between the continuum of frequencies of audible sound with harmonic relationships between notes and the spectrum of colours and the way they combine to make other colours or how colours harmonise or clash. (The word chromatic in music comes from the Greek word for colour.) The French mathematician Louis Bertrand Castel, following theories of Kircher (1646) and Newton (1704), invented an optical harpsichord in 1725 which displayed a different colour with each note played. Bach's contemporary Telemann went to see it in 1738, composed pieces for it and wrote about it Klotz (2014). Goethe mentioned some parallels between colour and music in his Farbenlehre in 1810, for example relating powerful graphic art to major keys and gentle with minor keys. The nineteenth century Scottish artist David Ramsay Hay made connections between Pythagorean proportions in music and colours, architectural proportions and the human body Hay (1838). Figure 3.3 from 1838 shows an example of relating the major triad to the three primary colours and the seven notes of the scale to other colours mixed from these as well as relating the notes of the musical triad to the geometric shapes circle, triangle and square (Internet12).

One of the most recent and thorough representations of music by colours is the work "Goldberg Variations 30 + 2" by Benjamin Samuel Koren, exhibited at the Bach Museum in Eisenach, Bach's birthplace (Fig. 3.4). This uses a sophisticated algorithm to convert the MIDI format of the music into colours. Each row represents a variation and the colours are mixed from the combination of notes played. The blue G Major columns at the beginnings and endings, and the golden D Major cadences in the middle can clearly be seen. A companion work on Beethoven's Diabelli variations also encodes the dynamics and attack and decay of each note on the fortepiano as colour saturations. See Koren (2015) and (Internet13) for a fuller description.

3.17 Beyond Bach

Fig. 3.3 Hay's relation of notes and colours. From Hay (1838) Public domain archive.org. Digital image courtesy of Getty's Open Content Program

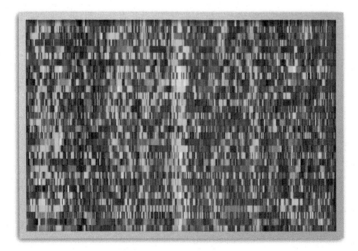

Fig. 3.4 Goldberg 30 + 2 by Benjamin Samuel. Goldberg Variation 30 + 12 (2010) © Benjamin Samuel Koren

3.17.5 Other Puzzles

Edward Elgar's Enigma Variations are also a well-known puzzle where the theme is hidden, and various numerical solutions have been proposed (Internet14).

3.18 Combined Codings

Some of the above could be combined, e.g. Prautzsch (2000) mentions the bar number and the number of notes in that bar to give the chapter and verse "coordinates" of a Bible reference. However, this does not indicate which book of the Bible should be used and there is nothing to indicate where the message is in a composition.

3.19 A Cryptographic Example

An interesting variant is given by (Licht 2009), winner of a youth science slam in Münster, Germany in 2011. She uses a combination of the notes of the scale (Sect. 3.3) and two note lengths, long and short (Sect. 3.5) to encode the letters of the German alphabet, the numbers and some punctuation marks, 45 symbols in all. As was pointed out in Sect. 3.3, using the notes would not give a pleasant melody. Licht overcomes this by analysing the frequencies of the occurrences of notes in compositions by Bach, Vivaldi and Mozart and allocating the notes to the letters of the alphabet according to the frequency of the occurrences of the letters in language texts (which, as we saw in Sect. 2.1 is incidentally a measure of the information content).

Figure 3.5[4] shows the coding table. The upper row gives the text characters—the first character LZ is the space (Leerzeichen). The lower row gives the corresponding musical notes with the octave number and an "L" for a long note. The space as the most frequently occurring symbol in text is given the note d2 short, the most frequently occurring note in the analysed compositions, and the next most frequently occurring symbol "e" is given the next most frequent note e2 short, etc.

Licht also thought of the problem of denoting the start and end of the message if it is concealed in a composition and uses four consecutive short e2 notes to denote the start and end of the message.

(Internet15) shows some seventeenth and eighteenth century methods of encrypting messages in music, including (Wilkins 1694) with a more primitive version of the above and Thicknesse (1772) who advocates filling out the code with other notes to produce realistic music and using the direction of the tails of the notes to indicate which are part of the message, and (Bücking and Heinrich 1804) with a method of representing each letter by a musical phrase, as well a as an encryption table by Michael Haydn from 1808. Other methods of musical cryptography are given in Tatlow (1991).

The mathematical foundations of cryptography were laid by Claude Shannon, whom we met above, during the Second World War.

---

[4] The coding for full stop "h = l" is a misprint and should read "h0L".

3.20 Summary

LZ	a	b	c	d	e	f	g	h	i	j	k	l	m	n
d2	a2	e2L	c3	h2	e2	g2L	d1	g1	f2	a1L	h0	e1	d3	c2

o	p	q	r	s	t	u	v	w	x	y	z	ä	ö	ü
c1	a0	h1L	h1	g2	a1	f1	a2L	d2L	f2L	c2L	e3	a0L	g0L	a3

0	1	2	3	4	5	6	7	8	9	ß	.	,	!	?
h2L	g1L	d1L	d3L	c3L	e1L	g0	f3	f1L	c1L	e3L	h=1	g3	f3L	g3L

Fig. 3.5 Licht's cryptographic coding table. From Licht (2009) with the kind permission of Deutsche Hochschulwerbung und-vertriebs GmbH

3.20 Summary

Table 3.2 summarises the above. All possibilities allow indirect coding, so this is not given explicitly in the table.

Table 3.2 Summary of possible codings

Sections	Coding	Size of symbol set	Possible length of message in piece	Possible length of message in composition	Restrictions
3.2	Number of bars	Arbitrary	1	= no. of movements	
3.3	Notes	35	Arbitrary	Arbitrary	
3.4	Intervals	7, 24 or more	Arbitrary	Arbitrary	
3.5	Note Lengths	7	Arbitrary	Arbitrary	Must add up to correct number of beats in bar
3.6	Number of Notes	Practically limited to tens	Arbitrary	Arbitrary	
3.7	Number of Pieces, Movements or Sections	1–68	1	1	
3.8	Sum of the G-Values of Notes	34	Arbitrary	Arbitrary	Length of message not visible
3.9	Key Signature	17	1, rarely more	= no. of movements	
	Key Signature (key of the piece with major or minor)	34	1, rarely more	= no. of movements	

(continued)

Table 3.2 (continued)

Sections	Coding	Size of symbol set	Possible length of message in piece	Possible length of message in composition	Restrictions
3.10	Accidentals	7	Arbitrary	Arbitrary	Not likely to result in good music
3.11	Occurrences of Words	Arbitrary	Arbitrary	Arbitrary	Only usable in choral music
3.12	Rests	8	Arbitrary	Arbitrary	Must add up to correct number of beats in bar
3.13	Time Signature	Approx. 16	1, rarely more	= no. of movements	
3.14	Figured Bass	39	Arbitrary	Arbitrary	Restricted by rules of harmony
3.15	Entries of a Theme	Practically limited to tens	1–2	= no. of movements	
3.17.1	BWV Numbers	1128	1	1 Several for a concert programme	Could not have been used by Bach
3.17.2	Frequencies	60 (for 5 octave instrument)	Arbitrary	Arbitrary	Could not have been used by Bach Not whole numbers
3.17.3	Morse Code	Full alphabet, letters, numbers, punctuation	Arbitrary	Arbitrary	Could not have been used by Bach

3.21 The Real Coding

It would be remiss not to consider the information encoded into a musical score for its intended use.

The basic information in a score is the pitch, duration and loudness of each note (This is also encoded in the Musical Instruments Digital Interface MIDI, as we shall see in Chap. 6.).

The pitches of the notes are modified, or further refined, by the key signature and accidentals, and even by the tuning and temperament. The pitch can also be varied on instruments such as the violin or human voice, e.g. with vibrato.

3.21 The Real Coding

The durations of the notes are modified by the time signature, markings such as dotting and staccato, as well as verbal instructions such as *andante* or *presto*. They can also be altered for expression, e.g. *tempo rubato*.

The other dimension is the relative loudness of each note, which is given by markings such as *forte* or *piano*, and partly given by the beat of the time signature. The ability to reproduce or "decode" this is dependent on the instrument, e.g. a harpsichord or organ can only change the loudness in steps by changing the manual or the registration (*Stufendynamik*).

An additional property of the sound produced when a note is played is the timbre, which depends on the overtones that come with the basic pure tone. This is determined by the instrument itself, e.g. violin, trumpet, horn, harpsichord, etc. or the combination of organ stops, and varies between instruments of the same kind. This is not usually notated in more detail than this, and for baroque keyboard music it is often not clear whether organ, harpsichord or clavichord is the intended instrument. The exact choice of organ stops is also rarely notated and stops with the same name will sound different on each instrument. The timbre can be slightly modified by the performer on some instruments, e.g. on strings by bowing or plucking nearer to or farther from the bridge. The sound is also affected by the acoustics of the room in which the instrument is played, especially the reverberation.

It is interesting to note that even the basic information is not given exactly in a musical score:

- The exact pitch of each note depends on the basis for the tuning, e.g. A = 440 Hz, and on the temperament to which the instrument is tuned, e.g. Kirnberger, Werckmeister, equal temperament, etc.
- The exact timing of each note is also flexible unless a metronome beat is given (e.g. ♪ = 80), and even then, will be given subtle variations by the performer (e.g. overdotting, rubato). The metronome was not invented until 1816, so there are no metronome markings in Bach's scores.[5]
- The exact loudness of each note is also a matter of interpretation by the performer as well as the limitations of the instrument.

This is one reason why MIDI files derived directly from a score and played on a digital instrument often sound bland and uninteresting—all the notes are exact in pitch, length and loudness, and the variability of human interpretation is missing. More on this in Sect. 6.1. Of course, MIDI files can be captured from a live performance as well, and these are less limited—only the loudness is limited to 256 different values.

MIDI recordings have mechanical predecessors for example piano rolls, similar in principle to the paper tape of early computers and teletypes, with rows of holes in a roll of paper or folded cardboard which cause a mechanism to strike the keys, or

---

[5]There is much discussion in the literature about *tempo giusto* or *tempo ordinario* concerning the correct tempo for baroque music. See also Sect. 10.9.

small musical boxes with a metal drum with protrusions that pluck the tuned metal comb that makes the sound. In the eighteenth century, there were also attempts to capture the nuances of live performance with mechanisms that marked a paper roll as the keys of a harpsichord were pressed (Klotz 2014). There is an ongoing project at the Musical Instrument Museum of the University of Leipzig to read piano rolls with a computer and store them in a standard format.

3.22 Notes for Researchers

> Anyone looking for possible codings should first ask themselves:
> - What is the maximum number of symbols that can be encoded, and are these sufficient to encode the required alphabet or other symbol set?
> - What is the maximum length of a message that can be encoded in the piece or work in question?
> - Can the presence of a message and its start and end be recognised?

> There are many ways of coding messages or references into music, some of which have apparently not been explored.

> As we shall see in Chap. 4, coding a word as a number (paragram or gematria) has such a huge amount of ambiguity that an attempt to decode a number will often give a chance result.

Chapter 4
Ambiguity in Decoding

4.1 Preamble

Most numerologists are primarily concerned with gematria or paragrams, i.e. the encoding of words using a numeric alphabet, and most of the methods given in Chap. 3 would be used in this way. We have shown in Sect. 2.5 that this form of hash coding has ambiguity when we come to decode the numbers back into words. As I learned from reading (Kramer 2000), the scientific term for this is "isopsephy" from the Greek meaning the same number of pebbles (used for counting, like an abacus).

The ambiguity was sometimes used intentionally to point out a relationship between two different words that have the same G-Value. Alternative, unusual, or even wrong spellings were sometimes used in much the same way as syllables are omitted from poems to fit into the meter, either as a last resort to obtain the desired number or as a hint to the reader to look for a paragram. Tatlow (2015) shows that this was common practice in poetry. Eighteenth-century German also had multiple correct spellings for many words as the medieval language gradually transformed into modern German.

We will now investigate the true extent of this ambiguity when trying to decode paragrams.

We consider the following three of the Latin alphabets as these are shown to be the most commonly used by Tatlow (1991):

- Latin Natural the letters are numbered sequentially 1, 2, 3, 4, ... 24.
- Latin Milesian the letters are numbered 1, 2, 3, ...9 then 10, 20, ...90, then 100, 200, ... 600.
- Latin Trigonal the letters are numbered with each step increasing by one 1, 3, 6, 10, 15, ... 300.

The full codings are given in Appendix D.1 An Excel macro to calculate G-Values is given in Appendix F.2.

4.2 Sources

To perform statistical analyses on the numeric codings of words, it would be desirable to have a dictionary of all words which were in use in Bach's time.

There are dictionaries from Bach's time referenced in Tatlow (1991) and Tatlow (2015) (Latin-German dictionaries Frisius 1704; Fritsch 1716), and later, e.g., Adelung 1781 or Grimm started in 1838, but although some of these are available online, they are not generally in a machine-readable form for computer analysis and in any case would not include all the variants of a word. It would also have been interesting to find a Latin dictionary or word list containing all variants (roots and endings) of Latin words, but again, none could be found.

Two approaches were taken. The first is to use a modern dictionary, which is available in a suitable form. The second is to use available historic resources, namely the Luther Bible and Bach's cantata texts, to obtain a list of common words from the time.

This could be extended as new sources become available.

4.3 Modern Dictionary

4.3.1 Method

The first dictionary used (Internet16) is an open-source German dictionary intended for use in simple spelling checker programs. This has the advantage of containing all the variants of a word, e.g. for the verb "kennen" (to know), "kenne", "kennst", "kennt", "kannte", "kannten", "kanntest", "gekannt", which would not normally be the case in a linguistic reference dictionary. At the time of the download, this contained 1,809,527 entries. Obviously, this contains very many words and abbreviations that would not have been in use in Bach's time (although the more esoteric enthusiasts may not find that a problem!).

The Umlauts ä, ö, ü and "scharfes S" ß are converted in the usual way to ae, oe, ue and ss respectively. Punctuation marks such as hyphens or full stops (for abbreviations) are omitted, i.e. counted as zero.

Non-German letters such as ë, Å, ç, ó, ô, ø, è, ê, ë, î, í, á, å á, â, à, û, ú, ñ, were counted as the unaccented letter.

The tables used are given in Appendix D.1.

There are also other differences which are not taken into account:

- Modern spellings, e.g. "nimmt" instead of "nimmet[1]"
- Poetic versions to omit a syllable, e.g. "Lob sei Gott sein'm eingen Sohn" instead of "Lob sei Gott seinem einzigen Sohn".[2]

---

[1] BWV 61 from Dürr (2010).
[2] BWV 62 ibid.

4.3 Modern Dictionary

As the 1.8 million words in the dictionary file exceeds the limit of the number of rows in an Excel table, the words were distributed over two worksheets, analysed with pivot tables and then combined.

4.3.2 Modern Dictionary with Latin Natural Coding

The dictionary was used to calculate the numeric equivalent of each word using the Latin Natural alphabet. An Excel function to do this is given in Appendix F.2.

The histogram (Fig. 4.1) showing the distribution of the number of words for each G-Value is almost a Gaussian normal distribution, but with a longer tail at the higher numbers. This is probably due to the German language having more longer words, e.g. by concatenating words to make a new composite word (e.g. the numbers are written as one word and can be arbitrarily long). The specificity of this to the German language has not yet been confirmed by doing a similar analysis on an English dictionary.

> **Histogram**
>
> A histogram shows the distribution of occurrences of a certain value. The horizontal axis gives the values and the vertical axis shows the number of times each value occurs. A random distribution, for example of the heights of adults, with height along the horizontal axis and the number of people of that height on the vertical axis, has the bell shape (called the normal or Gaussian

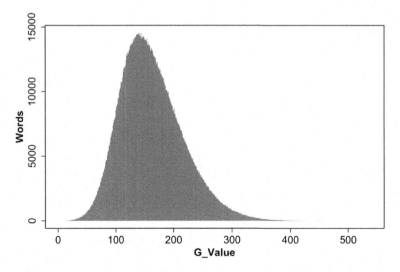

Fig. 4.1 Histogram of words per G-Value—modern German dictionary (Numeric Alphabet)

distribution), with most people centred around the average and tailing off at the extremes. There will be more on the shapes of distributions in Sect. 7.2.

Smend (1950) already admitted that "Of course every such word-number is ambiguous",[3] but he probably did not realise the true extent. The peak occurs at the G-Value 141, for which there are 14,549 possible words! On average, a number can represent 3597 words using this dictionary.

The number 14 (BACH) encodes 20 different words, none of them particularly meaningful.

The number 41 (JSBACH) encodes 306 different words.

4.3.3 Modern Dictionary with Latin Milesian and Trigonal Coding

The exercise was repeated on the modern dictionary for the Trigonal and Milesian alphabets—see Appendix D.1 for the tables. These were more commonly used than the Latin natural alphabet (Tatlow 1991). We would expect them to have fewer ambiguities in the coding, i.e. fewer possible words for the same number, because the use of tens and hundreds will spread the words further over the number space. It is interesting to speculate whether this may have been the reason that these were more widely used than the simple natural order alphabet. (Another possible advantage of these coding is given in Sect. 7.2.)

Figure 4.2 shows the distribution for the Latin Milesian alphabet variant 1 as given in Tatlow (1991) for the modern German dictionary. Variant 1 was used as it includes "W".

As expected, the spread is much wider (compare the numbering of the horizontal axis with that of Fig. 4.1), distributing the words over a larger range of numbers (see Fig. 4.4 for a direct comparison).

The number 530 has the highest number of words with 2252. The average number of words for each G-Value is reduced to 582. "BACH" = 14 now has 14 equivalent words in the Latin Milesian alphabet.

"JSBACH", 113 in the Latin Milesian coding, has 168 words with that value.

For the Trigonal variant 1, shown in Fig. 4.3:

"BACH" = 46 is equivalent to 6 words.

"JSBACH" = 262 has 59 one-word equivalents.

"JOHANN SEBASTIAN BACH" = 1103 has 13,561 one-word equivalents, but see Chap. 5 for the treatment of multiple words.

The maximum number of words with the same coding is 1710, and the average is 482.

---

[3]"Natürlich ist jede solche Wort-Zahl mehrdeutig" (Smend 1950) p. 25.

4.3 Modern Dictionary

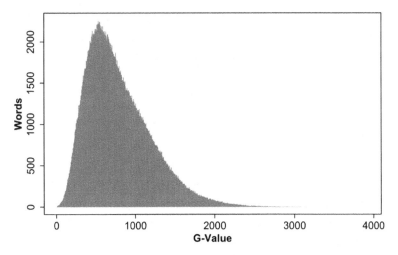

Fig. 4.2 Histogram of words per G-Value—modern German dictionary (Milesian Alphabet)

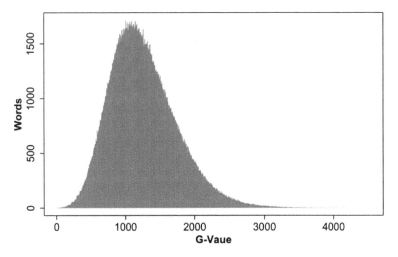

Fig. 4.3 Histogram of words per G-Value—modern German dictionary (Trigonal Alphabet)

These alphabets have the advantage of less ambiguity in the codings, but the larger resulting numbers make encoding in music more difficult (as observed by Rumsey (1997)—even the number of bars can get out of hand for words with letters towards the end of the alphabet.

A direct comparison of the three alphabets is shown in Fig. 4.4. The reduced ambiguity and wider spread over the number space of the Milesian and Trigonal alphabets can be clearly seen.

Note: although the Milesian alphabet uses a wider range ($Z = 600$) than the Trigonal ($Z = 300$), all the letters between B and T have larger values in the Trigonal,

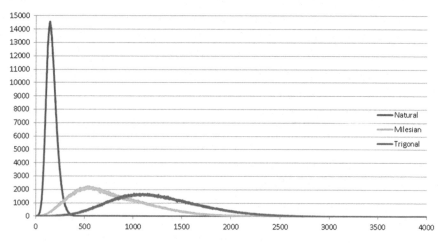

Fig. 4.4 Comparative histograms of words per G-Value—Modern German dictionary

so the Trigonal gives a larger spread of G-Values. See Appendix D.1 for the complete codings.

4.4 Historic Sources

4.4.1 Luther Bible

As said above, many words would not have existed in Bach's time. For comparison, the text of the Luther Bible was used, as this is a source available electronically and can be assumed to contain a representative sample of relevant words. This is about 160 years earlier than Bach, so would have a smaller vocabulary, and would not have used all the words available in the language even at that time. It also contains words no longer in the modern dictionary. The text of the Bible was taken from an available electronic source (Internet17)—this is a revision from the year 1912 so is probably not strictly the same vocabulary as the original, but we are more interested in the orders of magnitude than the actual words.

To use this, all the words were separated onto individual lines, sorted, duplicates removed and the G-Values calculated. This gives a vocabulary of 20,681 different words with the distribution shown in Fig. 4.5 using the Latin Natural alphabet. The shape of the distribution is very similar to that of the modern dictionary in Fig. 4.1 above. Here the G-Value 85 has the most equivalent words with 275.

The G-Value 14 has 6 words and 41 has 96 words.

On average a number can represent 74 different words from this compendium.

4.4 Historic Sources

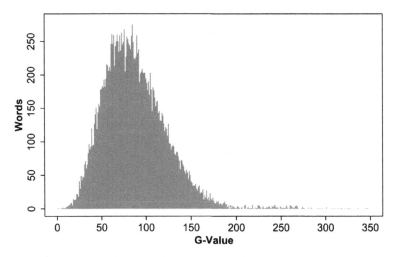

Fig. 4.5 Histogram of words per G-Value—German Luther Bible

4.4.2 Cantata Texts

A further source of historic words is the texts of Bach's cantatas. Most of the texts are available on the Internet, e.g. in (Internet18). Here the cantatas are each on individual Internet pages, but luckily there are index pages listing the links to the text pages. The index pages were manipulated with a sophisticated text editor[4] to create a batch file to automatically download the cantata texts. These were then converted to plain text[5] and copied into a single file. Using the text editor again, the individual words were split out and sorted, punctuation characters removed, some typing errors corrected and duplicates removed. This gives just over 12,000 words with an average G-Value of 55 and a maximum of 173 in the Latin Natural alphabet. The histogram is in Fig. 4.6.

4.4.3 Combining Historic Sources

Putting the words from the Luther Bible and the Bach Cantatas together gives a total of nearly 27,000 words. For Latin Natural coding there is an average of 94 words for each G-Value and a maximum of 347. This is shown in Fig. 4.7, which has a similar shape to the others.

---

[4] For the technically interested, using regular expressions in the TextPad editor—www.textpad.com.
[5] Using a macro in Microsoft Word to open each HTML file and store it as a text file.

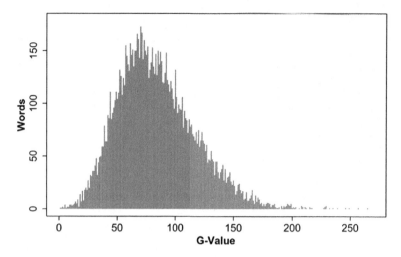

Fig. 4.6 Histogram of words per G-Value—Cantata texts

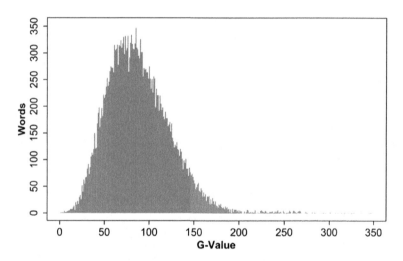

Fig. 4.7 Histogram of combined historic sources—Latin natural

The cantatas add one further word to those with the G-Value of 14 giving 7 words. The G-Value 41 has 122 words.

The effects of using the Latin Milesian (see Fig. 4.8) or Trigonal (see Fig. 4.9) alphabets are similar to the effects on the modern dictionary, as would be expected. A direct comparison is shown in Fig. 4.10.

The overall results are summarised in Table 4.1.

4.4 Historic Sources

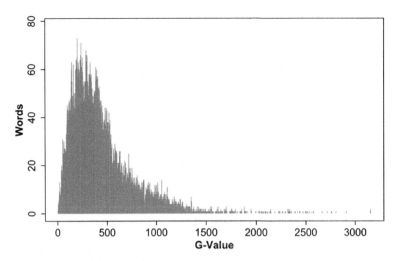

Fig. 4.8 Histogram of combined historic sources—Latin Milesian

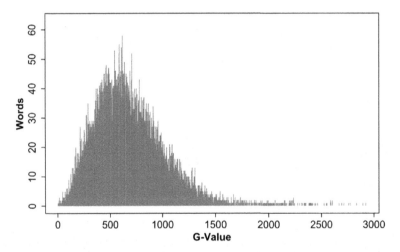

Fig. 4.9 Histogram of combined historic sources—Trigonal

Of course the unambiguous coding as Gödel numbers shown in Sect. 2.8 will spread the words out extending to huge numbers on the x-axis and mostly at 0 on the y-axis with occasional peaks of 1 as there are many numbers that do not encode any word, and no numbers that encode more than one word.

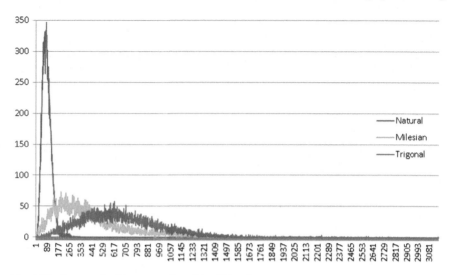

Fig. 4.10 Comparative histogram of combined historic sources

Table 4.1 Summary of dictionary statistics

Dictionary	Alphabet	Average number of words with same G-Value	Maximum number of words with same G-Values	Number of words coding as BACH	Number of words coding as JSBACH
Modern (1,809,527 words)	Latin Natural	3,597	14,549	20	306
	Latin Milesian 1	582	2,252	14	168
	Trigonal 1	482	1,710	6	59
Luther Bible and Bach Cantatas(26,994 words)	Latin Natural	94	347	7	122
	Latin Milesian 1	17	73	6	20
	Trigonal 1	15	58	4	5

4.5 Summary

Table 4.1 summarises the results, comparing the key values for the modern dictionary and contemporary sources with each of the three most common alphabets.

The large amount of ambiguity in the numerical coding of words lends support to the view that many supposed hidden messages are due to chance rather than intention. As we have seen, even Smend was aware that such numbers are ambiguous, and in the eighteenth century the ambiguity was sometimes intentionally used as part of the message.

4.5 Summary

Fig. 4.11 Loss of information

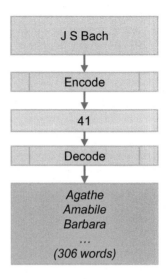

Figure 4.11 is a simplified reminder of how the encoding of words as single numbers loses information and makes it impossible to recover the original meaning in most cases (hashing as explained in Sect. 2.7).

As lower numbers have less ambiguity, the number 14 for BACH is at the lower end of the ambiguity scale.

4.6 Notes for Researchers

> Anyone looking for gematria should be aware of the large amount of ambiguity and be open to alternative meanings of any number, including references as well as direct coding.

> Bear in mind that there are many different encodings of the alphabet.

> If one is proposing that an author or composer intentionally created gematria, one must consider the historical practices and techniques of the time.

As with any speculative interpretation of intention, beware of the biases described in Chap. 12.

Chapter 5
Multiple Words and Partitioning

5.1 Partitioning and Permutations

Apart from representing a single word, a G-Value number may represent multiple words or a phrase, e.g. 158 = "JOHANN SEBASTIAN BACH" in the Latin natural alphabet. If we wish to decode a number as a G-Value for a phrase, we must first find all the possible sets of numbers that add up to this value. In mathematics, finding all the sets of numbers that sum up to the number in question is called partitioning—see the first box.

> **Partitioning**
> As a simple example, the number 5 can be partitioned in 7 different ways:
>
> $$5$$
> $$4 + 1$$
> $$3 + 2$$
> $$3 + 1 + 1$$
> $$2 + 2 + 1$$
> $$2 + 1 + 1 + 1$$
> $$1 + 1 + 1 + 1 + 1$$
>
> Note that this only includes unique sets, for example 4 and 1 but not the same numbers in a different order, 1 and 4.
>
> The number of partitions increases rapidly with the number being partitioned. There are 135 partitions of the number 14, and there are 88,751,778,802 (over 88 billion!) partitions of the number 158.

Furthermore, each set of numbers can be used in any order, so we must consider all permutations—see the second box.

Permutations

Permutations are different orderings of a set of items. For example, 3 items A, B, C can be ordered in 6 different ways:

$$A\ B\ C$$
$$A\ C\ B$$
$$B\ A\ C$$
$$B\ C\ A$$
$$C\ A\ B$$
$$C\ B\ A$$

4 items have 24 permutations,
5 items have 120 permutations.
The general formula for n items is n! (n factorial) or $n \times (n-1) \times (n-2) \times \ldots$, which for 4 items is $4 \times 3 \times 2 \times 1 = 24$.
This increases rapidly with the number of items.

An early example of the use of permutations is by Ramon Lull (1274) in the thirteenth century AD. Lull used permutations of symbolic words in an attempt to ascertain whether a statement was true. He even created a circular calculator to form the permutations (see Sect. 13.3).

Leibniz's dissertation *De arte combinatoria,* published in Leipzig in 1666, includes the formula n! for permutations as well as the mathematical technique of partitioning (Leibniz 1666). In 1677, Harsdörffer quoted various estimates for the number of ways the letters of the alphabet can be combined and in 1692 compared this to the size of the earth ((Schwenter 1692) Vol. 2 and Vol. 3).

5.2 Partitioning G-Values

To find all the possible phrases that could be denoted by a number such as 158, we must proceed as follows:

- Obtain all the partitions of the number,
- For each partition, obtain for each element the list of words that have that G-Value,
- Obtain all the combinations of these words,
- For each combination, obtain all the permutations.

A program was developed in Microsoft Excel to obtain the partitions. However, this soon exceeded the limits of the author's personal computer for larger numbers.

5.2 Partitioning G-Values

The partitions with lower numbers are of less interest, unless we are looking for names with initials (e.g. 1=A in A. Vivaldi), so the program was modified to allow a threshold to omit the partitions with numbers below that threshold. However, even with a threshold of 10, the partitions of 158 exceed the number of rows an Excel[1] spreadsheet can hold. A threshold of 11, i.e. excluding partitions which have numbers below 11, gives nearly a million (769,659) partitions of 158 using the modern dictionary.

An example is shown in Fig. 5.1:

Consider that each of the roughly 1 million partitions, or 88 million without the threshold, is a row of numbers, and below each number, you could show all the words which code to that number, which can be anything between 10 and 13,500 words, on average 3597 words (as we saw in Chap. 4), and then look for combinations of words

88 billion ways of partitioning the number 158:

158				
147	11			
146	12			
145	13			
144	14			
143	15			
142	16			
141	17			
140	18			
139	19			
138	20			
137	21			
136	22			
136	11	11		
...				
62	39	31	13	13
62	39	30	27	
62	39	30	16	11
62	39	30	15	12
62	39	30	14	13
62	39	29	28	

62 (1 569 words)	39 (303 words)	29 (123 words)	28 (100 words)
Aarauer	Abbacher	Abbild	abgehe
Abbauens	Abdallah	Abgas	agil
abbaufähig	abdiene	abhalf	Alban
Abbauloch	abdocke	abnage	alle
abbedingt	abebbende	affig	Anna
abbeizt	abfackle	Agape	Asi
Abbiegens	abfärbe	Akne	Aug
abblasende	abfeile	Alabama	babbele
abblendet	abfiele	Alice	Baby

Fig. 5.1 Partitions and multiple words

---

[1] Microsoft Excel is limited to 1,048,576 rows by 16,384 columns.

across the columns which would give a meaningful phrase in a certain permutation. A random example partition with four numbers, $62 + 39 + 29 + 28$ is shown. Using the modern dictionary from Sect. 4.3, we find that the numbers could decode to 1569, 303, 123 and 100 words, respectively. We therefore have $1569 \times 303 \times 123 \times 100 = 5,847,506,100$ possible combinations for this example alone.

The words can be used in any order, as the partitions exclude alternatives with the same numbers in different orders. We must therefore consider the number of permutations that are possible. The number of possible permutations of four items is $4 \times 3 \times 2 \times 1 = 24$. So we have $24 \times 5,847,506,100 = 140,340,146,400$ possible phrases (that is over 140 billion) for this particular sample of the 1 million partitions (or 88 billion if we did not have to restrict to the threshold above).

Figure 5.1 shows extracts from the corresponding table for the Latin natural alphabet using a modern German dictionary.

For any row, any permutation of the words will give 158.

These particular values give $1,569 \times 303 \times 123 \times 100 = 5,847,506,100$ phrases and all permutations of each row must be tried. Four words have 24 permutations giving 140,340,146,400 possibilities.

For example: "Amalia darf Bücher haben" (Amalia may have books—with reference to the Anna Amalia library in Weimar) or "Mache Kabbala aus Alphabet" (make Kabbala from alphabet).

This gives a huge potential for prospecting for hidden phrases with the codings in Chap. 3. Even with a smaller contemporary dictionary, there would be a huge number of possibilities.

Böß (2009) for example finds the following for 158:

TIEFES NACHSINNEN (deep thinking)
EIGENES NACHSINNEN (own or original thinking)
JSB BSJ PETSCHAFFT (JSB BSJ seal)
DAS MYSTERIUM (The mystery)
BASSO CONTINUO (Figured bass)
CHROMATISCHE FANTASIE UND FUGE D MOLL (2×158) (Chromatic fantasy and fugue in d minor)
EIN DOPPEL OPUS (A double work)
FATA MORGANA CLAVIER (Mirage keyboard)
TONDENKWUNDER (Tone thinking miracle)
EIN BACH DOPPELWERK (A Bach double work)
JS BACH UND DIE ZAHLEN (J S Bach and the numbers)
1 GLORIA STRETTA
SEQUITUR FUGA
DIE WAHRE LÖSUNG (The true solution)
IN DER STIMMUNG (In the tuning/temperament or mood)
DAS LETZTE INSTRUMENTAL WERK (2×158) (The last instrumental work)
JS BACH SCHLUESSEL (J S Bach key)
BACH UND BEACHTE KEIN ENDE (Bach and note no end).

5.3 Composers' Names

The following six baroque composers (from Djossa 2019) have the same G-Value as Johann Sebastian Bach in the Latin natural order alphabet (158):

Benedetto Ferrari
Orazio Benevoli
Carlo Agostino Badia
Francesco Geminiani
Geminiano Jacomelli
Johan Helmich Roman

A further example of ambiguity is given in (Maul 2016/17), where it is noted that the words "Walther" and "Gebhardus" have the same G-Value of 82.

In the Milesian alphabet with Johann Sebastian Bach = 500, there is one other full name: Johann Vierdanck and four surnames: Brixi, Reusner, Vallotti, Pasqualini

In the Trigonal alphabet with Johann Sebastian Bach = 1103, there is also one other: Wilhelm Friedemann Bach.

Using the surnames alone, there are no others with the equivalent of "Bach" (14, 14 and 46 in the respective alphabets).

5.4 Notes for Researchers

> Introducing partitioning for multiple words increases the ambiguity of G-Values astronomically.

> Partitioning theory shows that dividing a single number into parts to represent a phrase or name with several words gives a huge number of possibilities. It is therefore almost certain that something will be found.

Chapter 6
Score Analysis

6.1 The Method

Having established some methods for analysing the possibilities for encoding words, we turn our attention to the search for significant messages in the music (see Sects. 3.3 and 3.8).

A written or printed score is not conducive to computer analysis, but, as already mentioned in Sect. 3.21, there is an electronic format for representing musical performance on keyboards—the Musical Instruments Digital Interface MIDI (see (Internet20) for an overview). This was developed to allow an electronic musical keyboard to control other electronic instruments such as synthesisers. It registers which key on the keyboard is pressed and with what velocity (i.e. the dynamics) and for how long. The data is transmitted as a serial stream of bytes and can also be recorded in computer files. MIDI files of many works can be found in the public domain on the Internet.

MIDI files can be created manually and edited with other software, e.g. to give an exact rendition of the timings from the score. This can be useful for a performer to hear the exact timing of a difficult passage, but, as we mentioned in Sect. 3.21, tends to be rather sterile for normal listening, because a human performance contains subtle irregularities which constitute the musical expression. Indeed, this problem was foreseen by Johann Joachim Quantz in his 1752 treatise on playing the flute (Quantz 1752), where in section XII §11 (see Fig. 6.1) he says: "One could artificially prepare a musical machine, that could play certain pieces with such speed and precision that no human could equal, neither with the fingers nor the tongue. This would likely cause amazement, but it would never be moving to the spirit, and after hearing such a thing a few times and knowing its nature, the amazement would soon cease".[1]

As we shall see in Sect. 6.6, there are many other formats for representing music with computers. I used MIDI for this exercise because, after conversion to text, only

---

[1] Translation by the author.

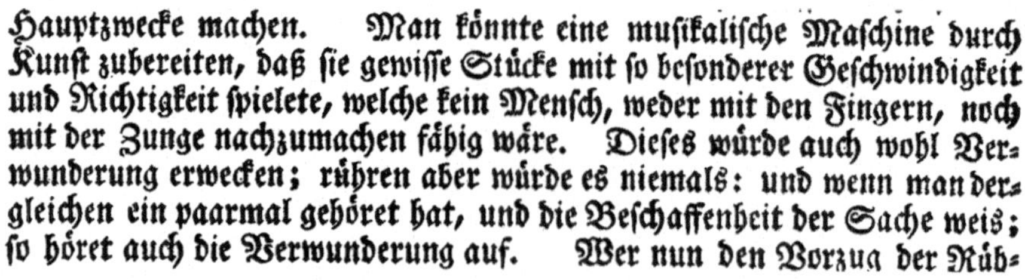

Hauptzwecke machen. Man könnte eine musikalische Maschine durch
Kunst zubereiten, daß sie gewisse Stücke mit so besonderer Geschwindigkeit
und Richtigkeit spielete, welche kein Mensch, weder mit den Fingern, noch
mit der Zunge nachzumachen fähig wäre. Dieses würde auch wohl Ver=
wunderung erwecken; rühren aber würde es niemals: und wenn man der=
gleichen ein paarmal gehöret hat, und die Beschaffenheit der Sache weis;
so höret auch die Verwunderung auf. Wer nun den Vorzug der Rüb=

Fig. 6.1 Quantz (1752) Section XII §11V. With kind permission of Bärenreiter-Verlag Karl Vötterle GmbH & Co. KG.

simple tools such as Excel and a text editor are needed, and MIDI files of music are widely available on the Internet.

MIDI files were converted to text using the program (Internet21). This conveniently gives for each bar the notes played and their duration.

The sample in Fig. 6.2 shows some extracts from the start of a MIDI file for the first movement (Adagio) of Violin Sonata No. 1 BWV 1001 with the important information in bold.

The lines begin with "BA" and the bar number and include "NT" with the note.

These files were processed with a text editor to reduce them to the desired data of the notes in each bar. A sample of the first bar of BWV 1001 is shown in Fig. 6.3.

This can easily be imported to Excel to count the notes in each bar and calculate the G-Values of the notes. Figure 6.4 shows the first bar of BWV 1001: in the left-hand screenshot column B is the bar number, column D is the note and column E is the calculated G-Value.

Excel pivot tables were then used to obtain the middle screenshot showing the number of notes in each of the 22 bars and the right-hand screenshot showing the sum of the G-Values for each bar with the grand total 4424.

The score of this bar from (Internet22) is shown in Fig. 6.5.

The reader should beware of the following potential pitfalls when exploring MIDI files:

- Sharps and flats should be converted to the German textual representation (A ♯ to Ais and A ♭ to As) as in Thoene (2016),
- Ignore the octave (e.g. A instead of A' or A-). This can be done by the G-Value calculation as the apostrophe, hyphen, etc., are given a G-Value of zero in the tables—see Appendix D.1.
- Convert B to H and B ♭ to B as in German notation (a procedure to avoid erroneously changing the B ♭ to H ♭ is: first change B ♭ to X, then change B to H, then change X to B),
- Rests may need to be removed, depending on how they are represented in the MIDI file.
- Accidentals are not always given correctly, for example a G ♯ may be given as A ♭, especially if the MIDI file was generated by performing on a keyboard—the

6.1 The Method

```
...
BA   1   CR          0   TR   0   CH  16   Time signature 4/4, clocks/mtick
24,  crotchets/32ndnote 8
BA   1   CR          0   TR   0   CH  16   Key F major
...
BA   1   CR          0   TR   1   CH  16   Text type 3: "Solo Violin"
...
BA   1   CR          0   TR   1   CH   1   NT  G'         1+1/8      von=100
...
BA   1   CR          0   TR   2   CH   2   NT  Bb              1     von=100
...
BA   1   CR          0   TR   3   CH   3   NT  D               1     von=100
...
BA   1   CR          0   TR   4   CH   4   NT  G-              1     von=100
...
BA   1   CR          0   TR   6   CH  16   Text type 3: "Sonata No. 1 in G
minor - BWV 1001"
...
BA   1   CR      1+1/8   TR   1   CH   1   NT  F'            1/8
BA   1   CR      1+1/4   TR   1   CH   1   NT  Eb'           1/8
BA   1   CR      1+3/8   TR   1   CH   1   NT  D'            1/8
BA   1   CR      1+1/2   TR   1   CH   1   NT  C'            1/8
BA   1   CR      1+5/8   TR   1   CH   1   NT  Bb            1/8
BA   1   CR      1+3/4   TR   1   CH   1   NT  A             1/8
BA   1   CR      1+7/8   TR   1   CH   1   NT  Bb           1/16
BA   1   CR     1+15/16  TR   1   CH   1   NT  G            1/16
BA   1   CR          2   TR   1   CH   1   NT  G             1/2
BA   1   CR          2   TR   2   CH   2   NT  C'              1
BA   1   CR          2   TR   4   CH   4   NT  A-              1
BA   1   CR      2+1/2   TR   1   CH   1   NT  F#            5/8
BA   1   CR      3+1/8   TR   1   CH   1   NT  E             1/8
BA   1   CR      3+1/4   TR   1   CH   1   NT  D             1/8
BA   1   CR      3+3/8   TR   1   CH   1   NT  E             1/8
BA   1   CR      3+1/2   TR   1   CH   1   NT  F#            1/8
BA   1   CR      3+5/8   TR   1   CH   1   NT  G             1/8
BA   1   CR      3+3/4   TR   1   CH   1   NT  A             1/8
BA   1   CR      3+7/8   TR   1   CH   1   NT  C'           1/16
BA   1   CR     3+15/16  TR   1   CH   1   NT  Bb           1/16
```

Fig. 6.2 MIDI example—Extract from start of BWV 1001

computer cannot easily tell in which key the current passage is and so may not choose the correct representation. See Meredith (2018) for solutions to this.
- Appoggiaturas may have been omitted or trills included in full (especially if the MIDI file was from a performance). The user must decide whether such notes should be included.
- Repeats may or may not be played, so would need to be removed or added, depending on whether one wished them to be included.
- It is also difficult to decide how to handle anacruses at the start of a piece and over repeats (e.g. the second movement of Partita 2 BWV 1002).
- There may simply be errors in the MIDI file.

When calculating the number of bars in a piece or the G-Values of the notes in the bars, it is important to decide how repeats, the first-time and second-time bars and anacruses are to be treated. More on this in Sect. 10.7.

BA	1	NT	G'	1+1/8	von=100
BA	1	NT	B	1	von=100
BA	1	NT	D	1	von=100
BA	1	NT	G-	1	von=100
BA	1	NT	F'	1/8	
BA	1	NT	Es'	1/8	
BA	1	NT	D'	1/8	
BA	1	NT	C'	1/8	
BA	1	NT	B	1/8	
BA	1	NT	A	1/8	
BA	1	NT	B	1/16	
BA	1	NT	G	1/16	
BA	1	NT	G	1/2	
BA	1	NT	C'	1	
BA	1	NT	A-	1	
BA	1	NT	Fis	5/8	
BA	1	NT	E	1/8	
BA	1	NT	D	1/8	
BA	1	NT	E	1/8	
BA	1	NT	Fis	1/8	
BA	1	NT	G	1/8	
BA	1	NT	A	1/8	
BA	1	NT	C'	1/16	
BA	1	NT	B	1/16	

Fig. 6.3 Processed MIDI data for first bar of BWV 1001

It is also possible to represent rests and time signatures in the output for exploring Sects. 3.12 and 3.13.

6.2 Counting Bars

When comparing results between (Tatlow 2015), (Thoene 2016) and the above score and MIDI files, there are some differences in numbers of bars in the movements—see Table 6.1. Thoene (2016) appears to count the anacrusis bars and does not consider how Bach counted. Tatlow (2015) is always careful to show how she is counting, considering anacruses, da capo, dal segno, repeats, time signature changes mid-piece, half bar lines, and counting at the breve.

The MIDI files (column 7) are all consistent with (Tatlow 2015) counting with repeats (column 2). (The * indicates where an initial anacrusis is not given a number in the score but does appear as bar 1 in the MIDI; an exception is the Ciaccona, where the initial anacrusis bar is numbered 1).

The numbers of bars in Tatlow (2015) without repeats (column 1) are consistent with the last bar number (column 5) in all cases except the Gavotte en Rondeau in Partita 3, which has an initial 8 bar repeat with anacrusis and then 90 bars with an anacrusis. (It should be pointed out that Tatlow is not looking for G-Values here).

6.2 Counting Bars

	A	B	C	D	E	F
1	BA	Bar	NT	Note	G-Value	Length
2	BA	1	NT	G'	7	1+1/8
3	BA	1	NT	B	2	1
4	BA	1	NT	D	4	1
5	BA	1	NT	G-	7	1
6	BA	1	NT	F'	6	1/8
7	BA	1	NT	Es'	23	1/8
8	BA	1	NT	D'	4	1/8
9	BA	1	NT	C'	3	1/8
10	BA	1	NT	B	2	1/8
11	BA	1	NT	A	1	1/8
12	BA	1	NT	B	2	1/16
13	BA	1	NT	G	7	1/16
14	BA	1	NT	G	7	1/2
15	BA	1	NT	C'	3	1
16	BA	1	NT	A-	1	1
17	BA	1	NT	Fis	33	5/8
18	BA	1	NT	E	5	1/8
19	BA	1	NT	D	4	1/8
20	BA	1	NT	E	5	1/8
21	BA	1	NT	Fis	33	1/8
22	BA	1	NT	G	7	1/8
23	BA	1	NT	A	1	1/8
24	BA	1	NT	C'	3	1/16
25	BA	1	NT	B	2	1/16

	A	B
1		Count
2	Row Label	of Note
3	1	24
4	2	24
5	3	31
6	4	17
7	5	22
8	6	18
9	7	21
10	8	23
11	9	25
12	10	32
13	11	27
14	12	20
15	13	30
16	14	27
17	15	23
18	16	28
19	17	23
20	18	29
21	19	21
22	20	18
23	21	33
24	22	4
25	Grand Total	520

	A	B
1		
2		
3	Row Label	Sum of G-Value
4	1	172
5	2	134
6	3	242
7	4	129
8	5	158
9	6	153
10	7	98
11	8	216
12	9	184
13	10	265
14	11	257
15	12	256
16	13	353
17	14	195
18	15	237
19	16	299
20	17	181
21	18	276
22	19	159
23	20	121
24	21	319
25	22	20
26	Grand To	4424

Fig. 6.4 MIDI data excel with notes in bar and G-values

Fig. 6.5 Score of first bar of BWV 1001. Staatsbibliothek zu Berlin—Preußischer Kulturbesitz. Mus.ms. Bach P 268. Public Domain 1.0. http://resolver.staatsbibliothek-berlin.de/SBB0001DAD600000000

Thoene (2016) (column 3) seems to have one bar more than the other sources, which is worrying, because both (Tatlow 2015) and Thoene (2016) use the numbers to show that Bach intentionally created proportional relationships and/or numerical references in these.

Thoene (2016) includes repeats in her lengths for Partita 2, but not for Sonata 1–3. (She does not treat Partitas 1 or 3).

Thoene (2016) has a slight inconsistency on p. 36, where for the Fuga of the third Sonata, two bars are used to obtain the 32 for the acrostic as opposed to only the first for all the other movements.

Table 6.1 Ways of counting bars in the Solo Violin Works BWV 1001–1006

	Movement A * represents an anacrusis and a + represents a different first/second-time repeat bar	1 Tatlow (2015)	2 Tatlow (2015) Repeats	3 Thoene (2016)	4 Thoene (2016) Repeats	5 (Internet22) score last bar no	6 (Internet22) score anacruses[2]	7 (Internet23) midi file
Sonata 1	Adagio	22	22	23		22	22	22
	Fuga allegro	94	94	95		94	94	94
	Siciliana	20	20	21		20	20	20
	Presto	136	272	137		136	136	272
Sonata 2	Grave	23	23	24		23	23	23
	Fuga	289	289	290		289	289	289
	Andante[3] + +	26	52	29		26	29	52
	Allegro	58	116	59		58	58	116
Sonata 3	Adagio	47	47	48		47	47	47
	Fuga *	354	354	355		354	353	355
	Largo	21	21	22		21	21	21
	Allegro assai	102	204	103		102	102	204

(continued)

[2]The number of complete written bars, counting an anacrusis and the corresponding final bar as one, and counting first-time and second-time bars separately, not counting repeats.

[3]Depends how the repeats are counted: bars 1-11a / 1-11b, 12 not repeated and 13-26a-27a / 13-26b. The score and the MIDI have 29 unique bars and a total of 52 bars are played.

6.2 Counting Bars

Table 6.1 (continued)

	Movement A * represents an anacrusis and a + represents a different first/second-time repeat bar	1 Tatlow (2015)	2 Tatlow (2015) Repeats	3 Thoene (2016)	4 Thoene (2016) Repeats	5 (Internet22) score last bar no	6 (Internet22) score anacruses	7 (Internet23) midi file
Partita 1	Allemande * ++	24	48			24	26	49
	Double * *	24	48			24	26	49
	Corrente * *	80	160			80	80	161
	Double Presto * *	80	160			80	80	161
	Sarabande +	32	64			32	33	64
	Double ++	32	64			32	34	64
	Tempo di Borea * *	68	136			68	68	137
	Double * *	68	136			68	68	137
Partita 2	Allemande * *	32	64		64	32	32	65
	Corrente * *	54	108		108	54	54	109
	Sarabanda +	29	52		52	29	30	52
	Giga * *	40	80		80	40	40	81
	Ciaccona *	257	257		256[4]	257	256[5]	257

(continued)

[4]Excludes initial anacrusis.
[5]The score counts the initial anacrusis as bar 1, unlike other similar occurrences.

Table 6.1 (continued)

	Movement A * represents an anacrusis and a + represents a different first/second-time repeat bar	1 Tatlow (2015)	2 Tatlow (2015) Repeats	3 Thoene (2016)	4 Thoene (2016) Repeats	5 (Internet22) score last bar no	6 (Internet22) score anacruses	7 (Internet23) midi file
Partita 3	Preludio	138	138			138	138	138
	Loure *	24	48			24	24	49
	Gavotte en Rondeau * *	92	108			100	98	109
	Menuett I + +	34	68			34	34	34[6]
	Menuett II + +	32	64			32	32	32
	Bourée * + * +	36	72			36	36	73
	Gigue * + * +	32	64			32	32	65

---

[6]Menuett I had the second repeat missing, and both Menuetts were in the same MIDI file—this was corrected manually.

6.2 Counting Bars 65

Although Thoene (2016) finds the names of Bach and Maria Barbara in the Ciaccona, these are taken from notes going across bars. Would Bach not have placed such codings cleanly within a bar to make it easier to recognise the start of the message?

6.3 Statistics

For our statistical analysis, I used the MIDI files including repeats and usually a separate bar for an initial anacrusis. Other anacruses were included in the corresponding final bar of the repeat with which they are played, giving a correctly complete bar. I also maintained a column for the bar numbers in the score, so that it was clear which bars are repeats.

For the statistics I combined the identifier of the movement and the bar numbers from the score and used this to remove the duplicate bars caused by the repeats.

Putting all the movements of the Sonatas and Partitas together, we see the following:

- For the sum of the G-Values in each bar there is a wide distribution as shown in Fig. 6.6, ranging between 1 and 31 occurrences.
- Of all the different G-Value sums, which fall in the range of 1–608, 260 numbers do not occur; any other of the 348 numbers can be found in at least one bar of the collection.

The most important Bach gematria numbers appear with the following frequencies in the grand total of 2423 unique bars (Table 6.2):

Taking pairs of consecutive bars, we find 1 occurrence of 29, 2 occurrences of 41, 14 occurrences of 81, 10 occurrences of 95 and 14 occurrences of 158.

To give another perspective, the following table gives the number of occurrences of the G-Values for words that Bach would certainly not have used (Table 6.3):

Table 6.2 Occurrences of "Bach" gematria in BWV 1001–1006

2	Bars with a sum of	14	Bach	A chance of	1 in 1212
10	Bars with a sum of	29	S D G or J S B	A chance of	1 in 242
18	Bars with a sum of	41	J S Bach	A chance of	1 in 135
18	Bars with a sum of	81	Maria Barbara	A chance of	1 in 135
12	Bars with a sum of	95	Maria Barbara Bach	A chance of	1 in 202
11	Bars with a sum of	158	Johann Sebastian Bach	A chance of	1 in 220

Table 6.3 Occurrences of "non-Bach" gematria in BWV 1001–1006

11	Bars with a sum of	87	Mozart	A chance of	1 in 220
22	Bars with a sum of	91	Beethoven	A chance of	1 in 110
13	Bars with a sum of	80	Shepherd	A chance of	1 in 186

– These have similar orders of magnitude to the above!

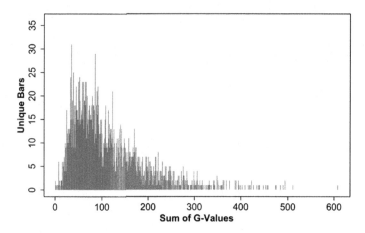

Fig. 6.6 Distribution of G-Values by bar for BWV 1001–1006 without repeats

(There will be more on the concept of "chance" in Sect. 7.2)

For the numbers of notes in each bar, the distribution is shown in Fig. 6.7.

The occurrences of gematria on the numbers of notes in a bar are shown in Table 6.4.

The maximum number of notes in a bar is 57, so higher numbers do not occur.

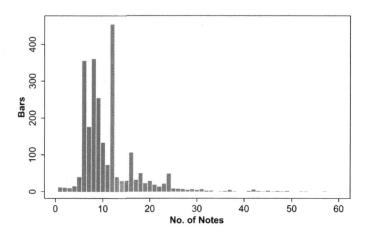

Fig. 6.7 Distribution of no. of notes in bar for BWV 1001–1006 without repeats

Table 6.4 "Bach" gematria by numbers of notes in BWV 1001–1006

30	Bars with a total of	14 notes	Bach	A chance of	1 in 81
7	Bars with a total of	29 notes	S D G or J S B	A chance of	1 in 346
2	Bars with a total of	41 notes	J S Bach	A chance of	1 in 1,212

6.3 Statistics

Table 6.5 Occurrences of "Maria Barbara Bach" = 95 in BWV 1001–1006

Movement	Bar in MIDI	Bar in score	No. of notes	Sum of G-Values
P1_6 double of Sarabande	7	7	9	95
P1_7 tempo di Borea	11	10	8	95
P1_8 double of tempo di borea	11	10	8	95
	61	40	8	95
	84	63	8	95
P2_2 corrente	70	45	9	95
P2_3 sarabanda	18	10	13	95
P2_5 ciaconna	239	239	18	95
S1_2 fuga allegro	76	76	15	95
S2_2 fuga	106	106	10	95
S2_3 andante	38	38	16	95
S3_2 fuga	96	95	8	95
No. of occurrences	**12**			

Looking at the G-Value 95 for Maria Barbara Bach, and considering the theory in Thoene (2016) that the work, or at least the Ciaccona is a tombeau for her, we see that it occurs in the collection as shown in Table 6.5. Here I have included a column for the bar numbers in the score—only one of the 12 unique occurrences is in this movement. Thoene (2016) finds most occurrences over bar boundaries, not as complete bars.

Doing the same for 81 (Maria Barbara) gives similar results—see Table 6.6, with 18 unique occurrences, two of these in the Ciaccona.

Tables 6.7, 6.8, 6.9 and 6.10 give the results for the other words and phrases showing that those that Bach could not have included occur in a similar pattern to the ones he could have, with "Beethoven" occurring most frequently.

6.4 Further Applications

Once a MIDI file has been converted as above, it could be manipulated for other purposes, for example to search for occurrences of melodic phrases such as the B A C H motif.

It would also be easy to convert the notes into intervals, e.g. using + 1 for up a semitone and -1 for down a semitone, the ascending major scale would be represented as + 2 + 2 + 1 + 2 + 2 + 2 + 1. Searching for a motif in this representation would then also find transposed instances. There are other sources for this, in particular the Parsons code used in (Internet7) which we saw in Sect. 3.4.

This could be applied to investigate Thoene's claim that Bach wove chorale melodies into the Ciaccona.

Table 6.6 Occurrences of "Maria Barbara" = 81 in BWV 1001–1006

Movement	Bar in MIDI	Bar in score	No. of notes	Sum of G-values
P1_3 corrente	5	4	6	81
	98	65	6	81
P1_4 double of Corrente	78	45	12	81
P1_7 double of tempo di borea	12	11	7	81
P2_1 allemanda	2	1	13	81
P2_2 corrente	74	49	8	81
P2_3 sarabamda	21	13	8	81
P2_5 ciaconna	37	37	12	81
	139	139	10	81
P3_3 gavotte en rondeau	101	92	4	81
S1_2 fuga allegro	42	42	16	81
S2_2 fuga	54	54	11	81
	56	56	11	81
	78	78	7	81
	233	233	7	81
S3_2 fuga	155	154	10	81
	244	244	11	81
S3_4 allegro assai	74	74	12	81
No. of occurrences	**18**			

Table 6.7 Occurrences of "Johann Sebastian Bach" = 158 in BWV 1001–1006

Movement	Bar in MIDI	Bar in score	No. of notes	Sum of G-values
P1_3 corrente	66	33	6	158
P1_4 double of corrente	18	17	12	158
	92	59	12	158
P1_6 double of sarabande	18	10	9	158
P2_1 allemanda	3	2	18	158
P2_4 giga	59	38	24	158
P3_6 bourée	7	6	8	158
	9	8	8	158
P3_7 gigue	2	1	9	158
S1_1 adagio	5	5	22	158
S1_2 fuga allegro	94	94	15	158
No. of occurrences	**11**			

6.4 Further Applications

Table 6.8 Occurrences of "Mozart" = 87 in BWV 1001–1006

Movement	Bar in MIDI	Bar in score	No. of notes	Sum of G-Values
P1_7	55	34	7	87
P2_3	22	14	10	87
P2_4	13	12	24	87
P3_1	64	64	12	87
P3_1	66	66	12	87
S2_2	6	6	7	87
S2_2	29	29	7	87
S2_2	215	215	8	87
S3_3	4	4	19	87
S3_4	67	67	12	87
S3_4	95	95	12	87
No. of occurrences	**11**			

Table 6.9 Occurrences of "Beethoven" = 91 in BWV 1001–1006

Movement	Bar in MIDI	Bar in score	No. of notes	Sum of G-values
P1_3	29	28	6	91
P1_3	95	62	6	91
P1_4	76	43	12	91
P1_7	51	30	7	91
P1_8	21	20	8	91
P2_1	17	16	13	91
P2_3	23	15	13	91
P2_5	46	46	12	91
P2_5	168	168	15	91
P3_1	1	1	6	91
S2_2	72	72	9	91
S2_2	164	164	9	91
S2_2	217	217	8	91
S2_3	4	4	16	91
S2_3	6	6	19	91
S2_4	43	43	18	91
S2_4	44	44	19	91
S2_4	48	48	20	91
S3_2	6	5	7	91
S3_2	69	68	10	91
S3_2	154	153	9	91
S3_2	284	283	12	91
No. of occurrences	**22**			

Table 6.10 Occurrences of "Shepherd" = 80 in BWV 1001–1006

Movement	Bar in MIDI	Bar in score	No. of notes	Sum of G-values
P1_5	2	2	6	80
P1_5	22	14	11	80
P1_7	9	8	8	80
P1_7	83	62	5	80
P2_2	69	44	8	80
P2_5	8	8	12	80
P2_5	45	45	12	80
S1_2	64	64	16	80
S2_2	108	108	10	80
S2_2	192	192	8	80
S2_4	7	7	16	80
S3_2	104	103	10	80
S3_3	3	3	20	80
No. of occurrences	13			

6.5 Summary

Words that Bach certainly would not have hidden as gematria in the notes of the score occur just as frequently as words that he may have hidden. They also occur in many other movements of the six Sonatas and Partitas, not only in the Ciaconna as shown by Thoene (2016). One can find almost anything in this way and words that Bach certainly did not encode are just as frequent as those that (Thoene 2016) claims. Of course not being able to prove that Bach did place these messages is not the same as proving that he did not, as we shall see in Sect. 7.3 on hypothesis testing.

Figure 6.8 shows how the notes from the score are encoded using the example of Sonata No. 1 for violin solo BWV 1001, the 25th bar of the 4th movement. (The piece is in G minor so the Bs are flat, but B-flat is B in German—see the footnote on page 20).

An interesting corollary to the above is given in (Bar-Natan and McKay 1997) where they refute a claim that the book of Genesis in Hebrew contains prophetic names and dates coded in Equidistant Letter Sequences[7] by showing that a similar number of "prophecies" can be found in Tolstoy's "War and Peace".

---

[7] The dates are found by taking individual letters at regular intervals ("equidistant"), e.g. every fifth letter.

Fig. 6.8 Example Encoding Procedure

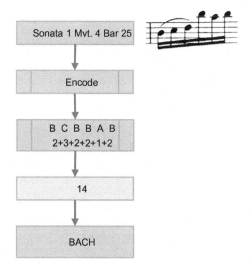

6.6 Other Representations and Tools

The above examples use the MIDI format as it is widespread, but there are more advanced formats being increasingly used by musicologists and musicians, some of which could probably do the above more easily and enable more sophisticated analyses.

There is a wide variety of representation formats and computer programs to process them for various purposes, e.g. printing scores for performance or publication, sounding the music on electronic instruments, musical analysis and converting and exchanging data between different incompatible systems. Some examples are:

Humdrum (Internet24)—a sophisticated set of command line tools for UNIX computers that could be used for the above analysis but would require UNIX skills.

Kern—an encoding of music used by Humdrum.

Music Encoding Initiative MEI—an encoding of music for the purpose of analysis and interchange, based on XML—see (Internet25). XML is a generic standard for mark-up languages originally for Internet content. As a specific example, all Internet browsers use HTML hypertext mark-up language.

Digital Alternate Representation of Musical Scores DARMS—intended for representation of the score.

MuseData—"to represent the logical content of musical scores in a software-neutral fashion"—see (Internet26), a general purpose representation intended to overcome the shortcomings of other formats.

MusicXML—an XML representation of scores.

Music21—"an object-oriented toolkit for analyzing, searching and transforming music in symbolic (score-based) forms" (Internet27).

Sibelius, Finale, MuseScore, SCORE—programs for desktop publishing of scores.

More information can be found in Selfridge-Field (1997) and (Internet26).

6.7 Notes for Researchers

> It is essential to decide on a clear and consistent way of counting bars, if this is relevant, including the treatment of repeats, anacruses, time signature changes, etc.

> The representation format may have ambiguities and errors.

> Consider the occurrences of any findings against the occurrences of random results (see also Sect. 7.3).

> When looking for gematria coded in the notes of a score, words that the composer definitely did not place there are found just as often as those that some authors claim he did.

Chapter 7
Statistical Methods

7.1 Preamble

We have already used histograms (Sect. 4.3.2) and talked about the chances of an event or observation (Sect. 6.3). Before examining the next subject, we need to introduce some more methods from statistics.

The main question we ask ourselves is: To what extent can the numerical properties in Bach's works be attributable to chance? Or to put it another way: How sure can we be that they were or were not intentional?

For considering this question various statistical terms and techniques spring to mind, e.g. probability, hypothesis testing, statistical significance, p-value, confidence interval, Bayes theorem. In this chapter, I attempt to briefly and simply introduce these, and show to what extent they are applicable to our questions and to point out some potential problems.[1]

Two important concepts in statistics are **population** and **sample**. The population is the total set of objects in which we are interested. Because we cannot usually measure or examine all members of the population, we take a representative sample and then use statistical methods to infer some property of the population from the examination of the sample. For example, if our population is a jar of 1000 sweets coloured red or blue, and we wish to estimate how many are of each colour without examining all the sweets, we can mix them up and take a random sample of 10 or 20 sweets, check their colours and infer the colours of the overall population. The larger the sample, the more accurate our knowledge of the population will be, but it will only be exact if we sample the entire population. Statistics can also gauge the degree of uncertainty in the estimate.

Many people have difficulty understanding statistical concepts, and there are many pitfalls, counter-intuitive facts and even differences of opinion between statisticians. There has also been much misuse and misinterpretation of statistics, both intentional and unintentional, as well as human biases (see Chap. 12).

---

[1] This is mostly taken from Bernstein (1996), Daniel (1998), Fenton (2013), Gullberg (1997), Sterne (2001).

Chance, proportions, probability, permutations, combinatorics and statistics have a long history dating back about 5000 years. The ancient Egyptians, Greeks and Romans used knuckle bones as dice and so already had some idea of the chances of throwing the winning cast as early as 3500 BC.

In 1494 Paccioli summarised the knowledge of arithmetic, geometry and proportions, including the solution to the problem of how to divide the stakes in a game of chance if the game is stopped before its conclusion.

In 1662 Graunt gathered and analysed statistics on mortality and other social data and introduced the concept of sampling—drawing conclusions about the entire population from a small random sample.

In 1696 Lloyds list was started to spread the high risks involved in shipping by means of insurance which involves raising premiums commensurate with the probability of accidents and the financial impact of the resulting loss.

All of this progress occurred before or during Bach's lifetime.

Continuing into the second half of the eighteenth and through the nineteenth century, the English church minister Thomas Bayes' theorem for inverse probabilities was published posthumously in 1764, Laplace formulated the central limit theorem in 1809, Gauss made the normal distribution famous, and statistics were increasingly applied in sociology with Quetelet's work on frequency distributions and the "average man" in 1849. Regression to the mean, tests for the goodness of fit of data to mathematical curves and the concept of statistical significance all came towards the end of the nineteenth century.

In the twentieth century we find progress on small samples, and the widespread use of statistical significance.

The advent of computers lead to the development of Monte Carlo simulation in the 1930s and 1940s.

See Appendix C for a summary of the history.

7.2 Probability and Distributions

We should be clear about the numerical expressions of probability—the following are different ways of expressing the same chance or probability:

(1) Odds of 1 to 4.
(2) A chance of 1 in 5 (we already used this notation in Sect. 6.3).
(3) A 20% chance.
(4) A probability of 20%.
(5) A probability of 0.2.

The preferred way of denoting probabilities among mathematicians is option 5, giving probabilities as values between 0 (cannot happen) and 1 (will certainly happen). However here, as we are often dealing with small numbers, we will often use option 4 with the corresponding values between 0 and 100%.

7.2 Probability and Distributions

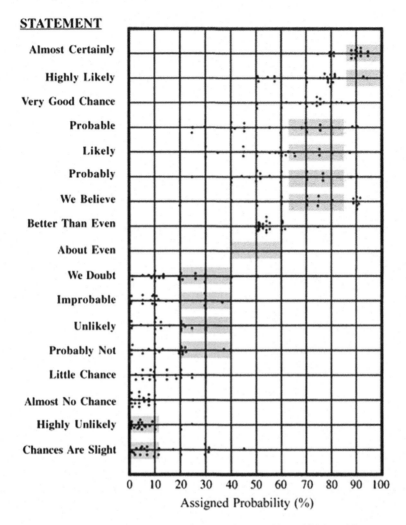

Fig. 7.1 NATO study of probability statements. From (Internet33) (public domain)

Different people can have very diverse subjective notions of whether a certain probability is for example highly unlikely, unlikely, likely, quite likely or highly likely. Figure 7.1 shows the result of a much quoted[2] study from 1970 of NATO officers, who were supposed to know how to interpret statements of likelihood in intelligence reports. For each statement down the left axis, the dots show the probabilities that the officers understood by it. The shaded areas show the range, proposed by leading CIA analyst Sherman Kent, for what each statement should be taken to mean. The spread of the dots is a clear indication that even experts will understand such terms in completely different ways.

---

[2]For example in (Internet33).

One thing that many people do not appreciate is that a probability of 0.5 or 50% means total uncertainty, the same as tossing a fair coin.

Hubbard (2009) quotes from a similar study on the interpretation of a climate change report that shows that people's interpretation of "very likely" ranges from 43 to 99% probability, and for "very unlikely" from 3 to 76%, even though they were given guidance on the meanings of the terms.

It is therefore best to completely avoid using such phrases.

Probabilities can be represented graphically as a distribution with the possible values of a variable plotted along the horizontal (x) axis and the probability or the number of occurrences of each value on the vertical (y) axis. The most common are the normal or Gaussian distribution, where the highest probability is the average value, falling off on either side to give the familiar bell-shaped curve, and the uniform distribution, where each value is equally probable. These are shown in Fig. 7.2. A distribution is similar to a histogram as described in Sect. 4.3.2, but there we had the actual values on the y-axis rather than the probabilities.

We can also consider the probability of multiple independent events all happening, which is given by multiplying the individual probabilities. To take a classical example, the probability of tossing heads with a fair coin is 0.5 and the probability of tossing three heads in a row is $0.5 \times 0.5 \times 0.5 = 0.125$. (This is the same as any other sequence of three results: there are 8 permutations of 3 symbols, so each has a probability of one eighth $= 0.125$.) This will be applied in Sect. 10.13.

Probability is important for information theory. We saw in Sect. 2.2 that the information contained in a received symbol is inversely proportional to the probability of it occurring at that point in the message.

We can also point out another advantage of the Latin Milesian alphabet over the Latin Natural alphabet for encoding messages. We saw in Sect. 2.4 that redundancy in the encoding can be used to detect errors and we mentioned the example of the parity bit. This has the effect that only half of the possible symbols are used—only those with an even number of ones in the binary representation are valid. The Latin Natural alphabet uses all the symbols 1–24 to represent the letters A–Z and has no redundancy if we ignore numbers above 24. The Latin Milesian alphabet uses the number space up to 600 (see Sect. 4.1 and Appendix D.1) and therefore has redundancy. A randomly generated symbol between 1 and 600 therefore has a

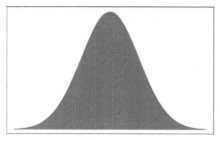

Fig. 7.2 Shapes of normal and uniform distributions

probability of 24/600 = 0.04 of being the code for a valid letter, so the probability of detecting an error which alters a symbol to a random value between 1 and 600 is 0.96 or 96%. For the Latin Natural alphabet, an error which alters a symbol to a random value between 1 and 24 will not be detected, as it will still be a valid symbol. Of course this logic does not apply to gematria, as the resulting G-Value can be any number.

Other important terms in this context are:

Mean—or the average, the sum of all the values divided by the number of values.

Standard Deviation—a measure of the spread of the values or how wide the distribution is.

Central Limit Theorem—taking the sums of samples of independent random numbers, the sums of the samples tend towards a normal distribution as more samples are added.

7.3 Hypothesis Testing and Significance

Classical hypothesis testing uses characteristics of sample distributions (the mean and standard deviation) to compare sample data sets, for example those taking a new drug and those taking a placebo, to decide whether the drug is effective, or taking samples from a manufacturing process to decide whether the occasional defective product is due to a systemic problem in the process or just due to chance variation.

In hypothesis testing the statistician takes a sceptical approach and formulates the hypothesis that the result of interest is due to chance and so would occur if the suspected cause were absent—this is called the "null hypothesis". Examples of null hypotheses in our case would be:

- The fact that a number we have found represents the numerical alphabet value of a certain word is due to chance (see Chap. 4).
- The fact that a number we have found represents the numerical alphabet value of a certain phrase is due to chance (see Chap. 5).
- A word found coded in a bar of a piece of music is due to chance (see Chap. 6).
- The fact that the works of a collection or the movements of a work can be divided up in a certain proportion is due to chance (we will come to this in Chap. 8).

In all these the "suspected cause" or the opposite of the null hypothesis is that features were intentionally put there.

The statistical significance or the p-value is the probability of (wrongly) rejecting the null hypothesis when it is in fact true, or put another way, the probability of observing the data assuming the null hypothesis is true. This gives the so-called statistical significance of the difference in sample data sets, e.g. the difference between those taking the drug and those taking the placebo. The use of statistical significance is a technique that is much disputed among statisticians.

The commonly used threshold value of 0.05 or 5% has been historically derived from statements [quoted from Cowles (1982)] such as:

> Vorausgesetzt nun, daß zwar ein großer Unterschied sey zwischen Erweislichkeit, *probabilitas*; Beweis, *probatio*; und Behauptung, *demonstratio*: da bey der ersten nur die Ursachen zu glauben stärker, als diejenigen, welche den Zweifel erregen; bey dem andern hergegen starke Ursachen zu glauben, und gar keine zu zweifeln vorhanden sind; bey der dritten aber das Gegentheil unmöglich dargethan werden kann:⁸ so soll dennoch bald erhel-

Fig. 7.3 From mattheson phthongologia systematica p. 15. From Niedersächsische Staats- und Universitätsbibliothek Göttingen (Göttingen State and University Library) ID PPN684732122 with permission

- "$p = 0.1$ (not very improbable), $p = 0.01$ (very improbable)—0.05 is the midpoint" (Pearson 1900).
- "three times the probable error in the normal curve would be considered significant" (Gosset, 1906),
- "30 to 1 as lowest odds for practical certainty that a difference is significant" (Wood and Stratton 1910)—approximately a probable error of 3.2.

These have no hard scientific basis, are subjective and, as shown in Sect. 7.2, will be interpreted very differently by different people.

This threshold of 0.05 is commonly used in sociology and medical trials, but other values are used in other fields. Particle physicists use 2.9×10^{-7} (1 in 3.5 million) as the threshold for announcing the discovery of a new particle, e.g. the Higgs boson—see Lyons (). There has been discussion as to whether this is too strict, but there have been cases where lower thresholds have led to announcements which were later proved false. Some fields in genetics use 5×10^{-8} (1 in 20 million)—see Fadista (2016).

There has been much discussion in the scientific community on the appropriate use and widespread misunderstanding of significance, e.g. (Daniel 1998), (Sterne 2001]. In March 2019 a complete supplementary issue of "The American Statistician" online journal was devoted to this—the reference is (Wasserstein 2019)—and this even warranted an online editorial in one of the leading scientific journals "Nature" (Internet28).

Figure 7.3 shows an interesting quote from Mattheson (1748) writing in 1748, which I found in Irwin (2015):

Now assuming a great difference between probability, *probabilitas*; proof, *probatio*; and assertion, *demonstratio*: since with the first only the reasons to believe are stronger than those that cause doubt; with the next on the other hand there are strong reasons to believe and none to doubt; but with the third the opposite cannot be considered possible.[3]

---

[3] Author's translation.

7.3 Hypothesis Testing and Significance

This is very reminiscent of the concept of hypothesis testing—putting it into more modern English with a slightly less literal translation and more emphasis on the Latin terms: a hypothesis can be considered probably true if the reasons to believe it are stronger than the reasons to doubt it; a hypothesis can be considered (tried and) proven if there are strong reasons to believe it and none to doubt it; a hypothesis can be considered demonstrably true if the opposite can be shown to be impossible.

Compare this to the statements from the beginning of the twentieth century, 150 years later, on hypothesis testing given above.

Popular Pitfalls:

- The p-value is sometimes interpreted as the chance that the null hypothesis is true given the data, whereas in fact it gives the chance of observing the data if the null hypothesis is true; these are fundamentally different, as we shall see in Sect. 7.6.
- A statistically significant result does not prove that the opposite of the null hypothesis is true. In fact, we can never prove the opposite of the null hypothesis, we can only give the probability of falsely rejecting the null hypothesis when it is actually true.
- Statistically significant does not necessarily mean important or significant in the real world—common sense has to be applied.

Applicability:

- We need a sample of multiple data points but here we only have one data point—a number we have found, a bar of music or a collection of works. However, we can apply the underlying concept, and do so in Sect. 9.2.

7.4 Confidence Interval

For a confidence interval we give a range within which we believe a value to be and the confidence level of that estimate.

A confidence interval of 95% means that, taking samples and dividing them into intervals determined by a set margin of error of $\pm e\%$, in the long run 95% of samples will contain the population value. Fenton (2013).

Popular Pitfalls:

- A confidence interval of 95% does not mean that there is a 95% probability that the value lies with the given range.

Applicability:

- We are not trying to make any statements of this nature, so this is not applicable here, but I mention it for completeness.

7.5 Monte Carlo Simulation

In Monte Carlo simulation, instead of taking or measuring samples from a population, one generates random numbers in a given range with a certain probability distribution. This is useful if one cannot obtain or measure samples but knows the range and distribution.

This is applicable here, for example for WTC1 we only have the one collection. We can generate random samples for a collection of 24 pieces, each with a random length in the same range as the lengths of the real collection. We want each length to be equally likely, i.e. a uniform distribution (not the familiar normal distribution with the Gaussian bell-shape).

> **Random or Pseudo-Random**
> In fact, a computer program cannot generate truly random numbers because it always uses some kind of algorithm or procedure, and they are therefore called pseudo-random numbers. Some generator programs are even intended to deliver a repeatable sequence of pseudo-random numbers. Truly random numbers can only be generated using special hardware—a recent advance in 2018 uses lasers and quantum theory to produce certified random bits (Internet29). However, a pseudo-random number generator is good enough for our purposes, as we shall see.

In statistics, the larger the sample size, the better the results, and a sample size of 100,000 is considered suitable. It is a common misconception that the sample size needed to obtain a meaningful result depends on the population size, in other words, the larger the number of things you are investigating, the larger the required sample. In the case of WTC1, the population is all possible combinations of 24 pieces with lengths between 53 and 159 bars. That is 107 different lengths that each of the 24 pieces can have—we allow many or even all the pieces to have the same length. The population is therefore 107^{24} or about 5×10^{48}. Our sample of 100,000 pieces (1×10^5) is really very small in comparison. But in fact, the confidence we can have in an estimate made from a sample depends on the sample size, and does not change much beyond a certain size, for example for a 95% confidence level and a margin of error of 5% the ideal sample size is 384 for populations of 1,000,000 and above.

We work with 100,000 samples firstly to obtain a good uniform distribution of the lengths from the (imperfect) pseudo-random number generator and secondly to ensure that our approximation to the complete population is adequate. This is demonstrated in Fig. 7.4 for the sort of data we will be using in the next chapters showing how many 1:1 proportions (solutions) there are for random samples of lengths of pieces. The figure shows a series of Monte Carlo simulations with the sample size increasing from 10 to 10 million. Down the left-hand side are the length histograms showing that the uniform distribution improves with increasing sample size, and down the right-hand side the corresponding distributions of the numbers of

7.5 Monte Carlo Simulation

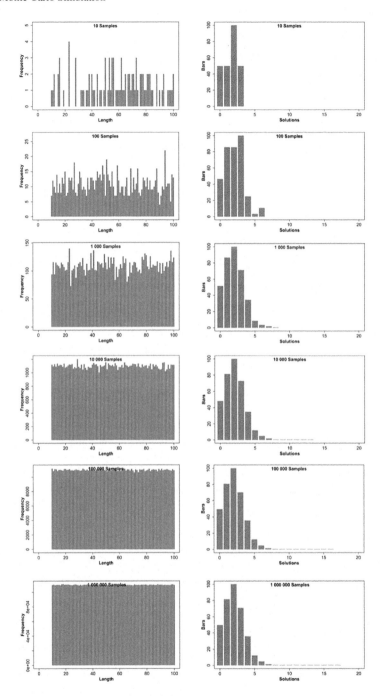

Fig. 7.4 Effect of increasing sample size

solutions. There is no difference between 1 and 10 million, and 100,000 is a good approximation. The number of samples we can use is limited by the time it takes to calculate the results for the samples.

The result histogram is steeper on the left and flatter on the right because in general, combinations with no solutions are more likely to occur than combinations with a large number of solutions. The exact shape depends on the number of pieces, the range of the lengths and the proportion.

The simulations are for 10 pieces with lengths between 10 and 100 bars.

7.6 Bayes Theorem

Bayes theorem and its extension into Bayesian networks enables initial estimates or prior beliefs to be combined to give a resulting probability distribution and then update the estimates as additional knowledge becomes available. Explaining the theory is beyond the scope of this book—see Fenton (2013) for a thorough treatment of the theory and usage, and (McGrayne 2014) for a history of its development and the surrounding controversies. But to summarise briefly, Bayes theorem states that the probability of A given B, i.e. the probability of A happening given that B has happened, is the probability of B given A times the probability of A divided by the probability of B, or in mathematical notation:

$$P(A|B) = \frac{P(B|A)P(A)}{P(B)}$$

The Bayesian approach of considering a probability given that a prior event has happened helps in avoiding some of the pitfalls pointed out in Sect. 7.3—with this mindset you cannot confuse the probability of a hypothesis being true given the data with the probability of observing the data given that the hypothesis is true. One classical example is that the probability of being pregnant given that the subject is female is not the same as the probability of being female given that the subject is pregnant. To demonstrate Bayes theorem with this, let us assume that half of the population is female, i.e. the probability of being female $P(F) = 0.5$, and that the probability of being pregnant (with child) given that one is female is 0.1, i.e. $P(C|F) = 0.1$. The overall probability of being pregnant $P(C)$[4] is the probability of being female times the probability of being pregnant, i.e. $0.5 \times 0.1 = 0.05$ as only the female half of the population can be pregnant. Then, applying Bayes theorem, the probability of being female given that one is with child is 1:

$$P(F|C) = \frac{P(C|F)P(F)}{P(C)}$$

---

[4] I use C for with Child rather than P for Pregnant to avoid confusion with the P for Probability.

7.6 Bayes Theorem

$$P(F|C) = \frac{0.1 \times 0.5}{0.05} = 1$$

The interactions between the priors (that which is given) and the resulting probabilities can be visualised as graphical networks—see Fig. 7.5 made with the AgenaRisk program in the free trial version as provided with (Fenton 2013). It shows the node "Pregnant" depending on the node "Female", and the values shown above are entered in the Node Probability Table for the "Pregnant" node, e.g. the probability of being pregnant given that being female is True is 0.1. The third screenshot shows the initial values. In the final screenshot it shows that if we enter pregnant as True the subject is certain to be female (probability = 1).

We can also test the reverse scenario and enter not being pregnant. This shows that a person who is not pregnant is slightly less likely to be female than male—if we take a random selection of people of both genders, some of the females will be pregnant, so more than half of those not pregnant will be male. This is slightly more complex to work out by hand, in particular the term $P(\sim C)$, the probability of not being pregnant, is the probability of being male (0.5) plus the probability of being female and not pregnant (0.5 × 0.9).

$$P(F|\sim C) = \frac{P(\sim C|F)P(F)}{P(\sim C)}$$

$$P(F|C) = \frac{0.9 \times 0.5}{0.5 + (0.5 \times 0.9)} = 0.474$$

This is shown in AgenaRisk in Fig. 7.6

Although the AgenaRisk program is primarily intended for the assessment of risks, the principles can be applied generally. In fact, many methods of risk management require that one also considers chances (in the sense of opportunities), treating them as negative risks.

Research shows that "access to a Bayesian network modelling tool, together with a limited amount of embedded training resources in its use, can help lay people solve complex probabilistic reasoning problems, involving multiple dependencies between uncertain pieces of information that can dynamically change as more information becomes available" (Cruz 2020).

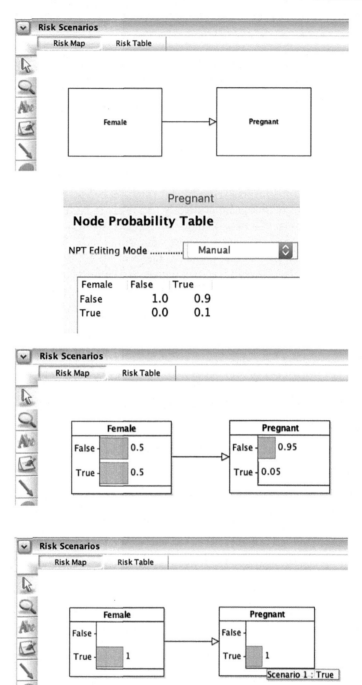

Fig. 7.5 Example Bayesian network. Screenshots from the AgenaRisk program with permission—see Fenton (2013).

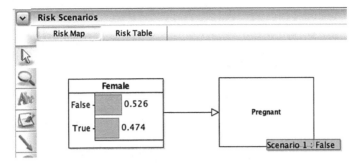

Fig. 7.6 Example Bayesian network—negative scenario

7.7 Notes for Researchers

> Probabilities are not intuitively understood by many people.
> Probabilities should always be stated as numbers between 0 and 1 or 0 and 100%.

> The use of Statistical Significance is disputed in scientific circles and should be used with caution and a full understanding.
> The value for declaring a result "significant" should be carefully defined before doing the research.
> The data must be presented openly in full so that the results can be reproduced by others.

> Monte Carlo simulation is very useful to widen the context of some findings. The number of samples simulated has a large effect on the distribution of the results up to a certain limit.

> Bayes theorem and Bayesian networks are very useful for understanding interactions between effects and for updating prior knowledge or assumptions on the strength of additional information.

Chapter 8
Exploring Proportions

8.1 Preamble

The discovery of the proportions that give rise to the musical scale and harmonies is attributed to Pythagoras, who lived around 500 BC. One legend has it that hearing blacksmiths forging with hammers of different weights he noticed that they produced harmonies or dissonances depending on the ratio of their weights. This is illustrated for example in the initial engraving in Athanasius Kircher's "Musurgia universalis" 1650 and in Quantz (1752) (Fig. 8.1).

Pythagoras is also credited with experimenting with dividing a string into different proportions with a monochord, and his ratios for the musical scale are shown in Raphael's School of Athens from 1511. Dividing the string into two equal parts for a 1:1 ratio gives the same note or unison, 1:2 gives the octave, 3:2 the perfect fifth, etc. The clavichord (see Chap. 1) can be seen as extending this, as the tangents divide the string to give the intended note, and older instruments used more than one tangent on the same string (fretting).

We do not wish to delve further into the subject of temperaments here—I mentioned them in Sects. 3.17.2 and 3.21 and will again in Sect. 13.2. For further reading, see for example Duffin (2007). Suffice it to say that the intervals between notes, which are adjusted by cents[1] to approach the goal of being able to play in all keys without unpleasant dissonances, might also be a source of desirable proportions. Some of these are given in Appendix E.

There are proportions which are not rational numbers, i.e. cannot be expressed as the ratio of two integers (whole numbers). Examples are the ratio of the diameter of a circle to its circumference π (pi) or the golden section which divides a line such that the ratio of the larger to the smaller part is the same as the ratio of the larger part to the whole. These are often approximated by a fraction, for example Metius' approximation to π 355/113, published in 1611, found by his father in 1585 [see Kramer (2000)], or the golden section can be approximated by the ratio of

---

[1] A cent is one hundredth of a semitone.

Fig. 8.1 Pythagoras and the Blacksmiths. From Quantz (1983, 2018) with kind permission of Bärenreiter-Verlag Karl Vötterle GmbH & Co. KG

two adjacent numbers from the Fibonacci sequence. These were not used for the proportioning of music in Bach's time (Tatlow 2006; Tatlow 2015, p. 105).

In the first century BC, Vitruvius, a Roman architect and engineer, wrote his ten books on architecture which include a description of the proportions of the human body in some detail: "the face, from the chin to the top of the forehead and lowermost roots of the hairline should be one-tenth of the total height of the body…" (Rowland), and a paragraph on relating the dimension of temples to the human body—the "Vitruvian Man".

> Similarly, indeed, the elements of holy temples should have dimensions for each individual part that agree with the full magnitude of the work. So, too, for example, the center and midpoint of the human body is, naturally, the navel. For if a person is imagined lying back with outstretched arms and feet within a circle whose center is at the navel, the fingers and toes will trace the circumference of this circle as they move about. But to whatever extent a circular scheme may be present in the body, a square design may also be discerned there. For if we measure from the soles of the feet to the crown of the head, and this measurement is compared with that of the outstretched hands, one discovers that this breadth equals the height, just as in areas which have been squared off by use of the set square.

There is no original drawing by Vitruvius himself, but several artists made drawings from Vitruvius' text in the fifteenth century, the best known interpretation being that by Leonardo da Vinci (Fig. 8.2). As Rowland (1999) points out, this is simplified,

8.1 Preamble

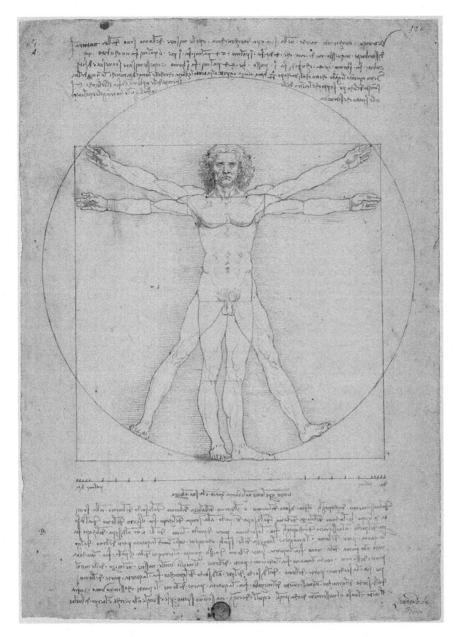

Fig. 8.2 Leonardo da Vinci's interpretation of Vitruvius. ©G.A.VE—Archivio fotografico foto: Matteo De Fina, 2019 *"su concessione del MIBACT—Gallerie dell'Accademia di Venezia*

as the arms and legs pivot from the shoulder and hip joints and not from the central navel, and Fig. 8.3 gives a more accurate but less elegant representation.

In the sixteenth century, Barbaro translated Vitruvius into Italian and Palladio wrote his four books on architecture and propagated the principles of simple proportions, firmly anchoring their use in architectural design. These have been a foundation for the arts ever since, and their use, for example in architecture, has endured through Palladio to Le Corbusier, who in 1940 defined "Le Modulor" as a unit of measure based on the proportions of the human body. Architects were still interested in the use of simple proportions, as witnessed by Wittkower (1998) in 1949. As we saw in Sect. 3.17.4, Castel in 1725, Goethe in 1810 and Hay(1838) in the 1830s also saw how the simple proportions applied to colours, architecture and drawings of the

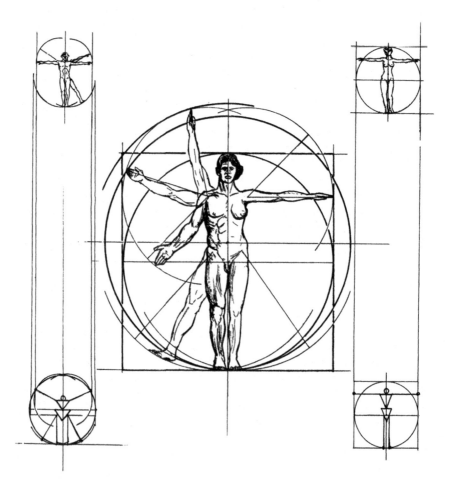

Fig. 8.3 Vitruvian man. From Rowland, Ingrid D. and Howe, Thomas Noble. Vitruvius Ten Books on Architecture. © Cambridge University Press, 1999. Reproduced with permission of the Licensor through PLSclear. Scanned by the author

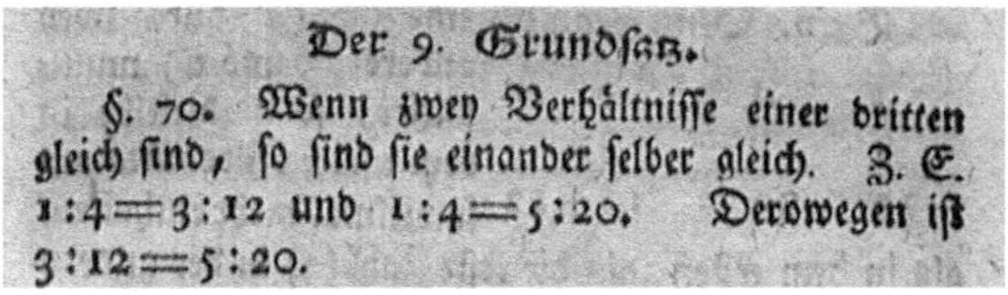

8.1 Preamble

> Der 9. Grundsatz.
> §. 70. Wenn zwey Verhältniffe einer dritten gleich sind, so sind sie einander selber gleich. Z. E. 1:4 = 3:12 und 1:4 = 5:20. Derowegen ist 3:12 = 5:20.

Fig. 8.4 Wolff Anfangsgründe Vol. 1 p. 61—Equivalence of Proportions. Digitalisat des Universitäts- und Landesbibliothek Sachsen-Anhalt in Halle (Saale); ID VD18 90,183,886. http://digitale.bibliothek.uni-halle.de/urn/urn:nbn:de:gbv:3:1-433575 Creative Commons 3.0 with permission

human body and how colour could be related to music. Architecture has also been called "frozen music" since about 1800 [see Pascha (2004)].

Tatlow (2015) points out that in Bach's time, proportions were not regarded as mathematics, but theology, the striving for the perfect unison of 1:1 or the next best 1:2. Of course, one does need some basic arithmetic to create proportions, but it was not done for mathematics' sake.

Christian Wolff's *Anfangsgründe aller mathematischen Wissenschaften* Wolff (1710) includes the arithmetic of proportions—see Fig. 8.4 for example and this shows that the associated mathematics were regarded as essential in the early 1700s:

> The 9th Principle §70. If two proportions are equal to a third, they are equal to each other. E.g. 1:4 = 3:12 and 1:4 = 5:20. Therefore 3:12 = 5:20.[2]

Proportional Parallelism was first introduced to musicology by Dr. Ruth Tatlow in Tatlow (2007) and developed in her ground-breaking book "Bach's Numbers – Compositional Proportion and Significance" Tatlow (2015). The book shows that many collections or works are structured to divide up into 1:1 and 1:2 proportions in multiple ways, e.g.

- Dividing a set of pieces into two subsets such that their lengths in bars are in proportion
- Dividing a set of pieces into two subsets such that the numbers of pieces in each subset are in proportion.

For example, as we shall see in Chap. 9, the Well Tempered Clavier Book 1 (WTC1) consists of 24 preludes and fugues. The pieces can be divided into two sets with a 1:1 proportion in the numbers of pieces as 24 preludes and 24 fugues or as 12 preludes and fugues in major keys and 12 in minor keys. Some examples of 1:1 proportions are shown in Fig. 8.5.

Further **layers** of proportion can be obtained by further subdividing the above according to the lengths in bars as shown in Fig. 8.6, or by the number of pieces. The collection has a total of 2088 bars. The 24 preludes and fugues can be divided into two sets of 1044 bars each, and these might each have 12 pieces to give a double parallel 1:1 proportion. Furthermore, each set of 1044 bars could be further subdivided into two sets of 522 bars each, etc.

---

[2] Author's translation.

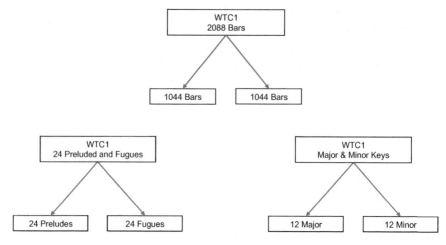

Fig. 8.5 Some 1:1 proportions in Well Tempered Clavier Book 1

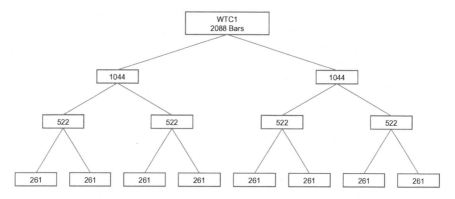

Fig. 8.6 Layers of proportion in numbers of bars

There are also possibilities to form proportions from the gematria of words in the titles of pieces—(Tatlow 2015, p. 232) gives a possible example for the Musical Offering.

A collection or work that exhibits proportions in two or more of these ways has **parallel proportions**. There may also be

- parallels between different collections
- parallels between the proportions within a work
- parallels between works within a collection
- parallels between sections of a piece.

The proportions in numbers of pieces are easy to find. Of course, the division of the WTC into equal numbers of preludes and fugues or equal numbers in major and minor keys lies in the nature of the collection and is not remarkable as a proportion.

8.1 Preamble

To find the ways of dividing up the pieces such that the numbers of bars are in a certain proportion is more complex and requires a computer program.

Yet another aspect of the aesthetics in all arts is symmetry, and as we shall see, this can also be applied here, for example if a proportion is made up of the first six pieces and the last six pieces it would form a symmetrical pattern. In its most perfect form, a symmetry would be combined with a 1:1 proportion.

We will start in Sect. 8.2 with the terminology and mathematics needed for finding the simple proportions in the numbers of bars and detecting patterns in these. In Sect. 8.3 we will extend this to cover multiple layers of proportions. Then, having summarised the new terminology in Sect. 8.4, we will get to know a computer program, the Proportional Parallelism Explorer, to explore these relationships—Appendix B contains the full user manual. The program is also suitable for exploring proportions beyond those looked at here.

Some sample results are given in Chap. 9 and Appendix A. The results are summarised in Chap. 10.

My aim is, where (Tatlow 2015) gives examples of parallel proportions in Bach's works derived with thorough historical methods, to use the computer to find all possible ways of obtaining a given proportion and explore the probabilities of these occurring by chance. The intention is primarily to provide some tools and to demonstrate the statistical and computer-based methods rather than draw any musicological conclusions.

8.2 Simple Proportions and Terminology

8.2.1 Sets and Pieces

In Chap. 1, we saw definitions for collection, work, movement and piece. The analysis of proportions can be applied at any level. There may be more than one way of dividing a collection into pieces. For example, WTC1 could be considered as 24 preludes and fugues (24 pieces) or as 24 preludes and 24 fugues (48 pieces). Or the six sonatas and partitas for solo violin could be taken as a collection of 6 works or a set of 32 pieces (counting the individual movements as pieces), or a single sonata from this collection could be taken as a work with the movements as its pieces.

For technical analysis of proportions, we do not need to differentiate these terms, so we shall simplify matters by just talking about proportions in the lengths of **pieces** in a **set**.

For proportional parallelism we are primarily interested in the numbers of bars in each piece and the total number of bars in the set, which we call the **lengths**. When determining the lengths of pieces, as we saw in Chap. 6, care is needed in treating anacruses, repeats, first and second repeat bars and da capo sections. It is recommended to follow the conventions used by Tatlow (2015).

8.2.2 Proportion

A **proportion** is a relationship between two numbers, also known as a ratio. They can be integer (whole number) relationships, e.g. 1:1 or 1:2, or irrational such as the relationship between the circumference and diameter of a circle π.

As (Tatlow 2015) has expounded, the use of simple whole number proportions has been a central tenet of aesthetics in all art forms since ancient times, with the most influence coming from the Greeks, in particular Pythagoras.

In proportional parallelism, Tatlow divides the set of pieces into two subsets such that their lengths in bars are in a certain proportion. A proportion of 1:1 divides the set into two subsets, each having an equal number of bars, a 1:2 proportion divides the set into two subsets where one subset has twice the number of bars as the other, etc.

Tatlow (2015) also considers having the number of pieces in the two subsets in the given proportion to give a double proportion.

To find which proportions are possible for a set of pieces we need to find the factors of the total length and then the partitions of each factor (see Sect. 5.1 for partitions). For example, the number of bars of WTC1, 2088, has the factors:

1, 2, 3, 4, 6, 8, 9, 12, 18, 24, 29, 36, 58, 72, 87, 116, 174, 232, 261, 348, 522, 696, 1044, 2088 which give the following proportions:

Table 8.1 Factors and proportions of 2088

Factor	Possible proportions
1	
2	1:1
3	1:2
4	1:3
6	1:5
8	1:7, 3:5
9	1:8, 2:7, 4:5
12	1:11, 5:7
etc.	

8.2 Simple Proportions and Terminology

Some proportions are not possible, for example 5 is not a factor of 2088 so there can be no 2:3 proportions. We always reduce the proportion to its simplest form, so for example proportion 2:2 for the factor 4 is reduced to 1:1 and not included separately.

Tatlow (2015) concentrates on 1:1 and 1:2 as these are the proportions used in Bach's time. The unity and 1:1 were considered perfection, and a number or proportion becomes less perfect the further away it is from this. 1:2 is the proportion of the musical octave, which is the most perfect harmony after the unison.

An Excel function to find the factors of a number is given in Appendix F.1.

Other sources of potential proportions were mentioned in Sect. 8.1.

8.2.3 Combinations

To find all the ways of dividing a set of pieces into a given proportion we need the mathematical concept of **combinations**, i.e. how many different ways can you select "k" objects from a set of "n" objects?

Combinations

The number of ways of getting "k" objects out of a set of "n" objects is represented by ^nC_k, for example $^{24}C_2$ denotes the combinations of any 2 out of 24. (^nC_k can be read as "from n pieces, take any k.")

For example, with $^5C_3$: from 5 pieces you can take 3 in ten different ways—there are ten combinations, $^5C_3=10$:

5, 4, 3	5, 4, 2	5, 4, 1
5, 3, 2	5, 3, 1	5, 2, 1
4, 3, 2	4, 3, 1	4, 2, 1
3, 2, 1		

The values can be obtained from mathematical tables, the Excel "COMBIN" function or scientific calculators.

The formula is $^nC_k = \frac{n!}{n!(n-k)!}$.

Here we need to go further and take the lengths in a set of pieces and find the combinations that add up to the appropriate fraction for a given proportion, e.g. for WTC1, which has a total of 2088 bars, we need:

1044 and 1044 bars for 1:1,
696 and 1392 for 1:2,
522 and 1566 for 1:3, etc.

Taking the example of 1:2, i.e. 696:1392, the proportion could be attained with:

1 of the 24 having 696 bars ($^{24}C_1$) and the other 23 having 1392,
2 of the 24 having 696 bars ($^{24}C_2$) and the other 22 having 1392,
...

22 of the 24 having 696 bars ($^{24}C_{22}$) and the other 2 having 1392 bars,
23 of the 24 having 696 bars ($^{24}C_{23}$) and the other 1 having 1392 bars.
So the total number of combinations we need to consider is:
$^{24}C_1 + {^{24}C_2} + \cdots + {^{24}C_{22}} + {^{24}C_{23}}$.
which is

24 + 276 + 2024 + 10,626 + 42,504 + 134,596 + 346,104 + 735,471 + 1,307,504 + 1,961,256 + 2,496,144 + 2,704,156 + 2,496,144 + 1,961,256 + 1,307,504 + 735,471 + 346,104 + 134,596 + 42,504 + 10,626 + 2024 + 276 + 24 = 16,777,214 (over 16 million).

The simple formula for this is $2^N - 1$, which for 24 is 16,777,215. The -1 is to exclude the case of taking 0 pieces $^{24}C_0$.

The 1:1 proportion is a special case.

Repeating the above exercise for the 1:1 proportion, we have:

1 of the 24 having 1044 bars ($^{24}C_1$) and the other 23 having 1044,

2 of the 24 having 1044 bars ($^{24}C_2$) and the other 22 having 1044,

...

11 of the 24 having 1044 bars ($^{24}C_{11}$) and the other 13 having 1044 bars,

12 of the 24 having 1044 bars ($^{24}C_{12}$) and the other 12 having 1044 bars.

Beyond this, we are just looking at the opposite arrangements, e.g. 13 of the 24 having 1044 bars ($^{24}C_{13}$) and the other 11 having 1044 bars, which is the opposite of $^{24}C_{11}$ which we already covered. Also, with 12 of the 24 ($^{24}C_{12}$), half of these combinations will be the opposites of the other half,[3] therefore we only consider half of the $^{24}C_{12}$.

So the total number of combinations we need to consider is:
$^{24}C_1 + {^{24}C_2} + {^{24}C_3} + {^{24}C_4} + {^{24}C_5} + {^{24}C_6} + {^{24}C_7} + {^{24}C_8} + {^{24}C_9} + {^{24}C_{10}} + {^{24}C_{11}} + {^{24}C_{12}}/2$.

which is

24 + 276 + 2024 + 10,626 + 42,504 + 134,596 + 346,104 + 735,471 + 1,307,504 + 1,961,256 + 2,496,144 + 2,704,156/2 = 8,388,607 (over 8 million, exactly half of the above general case). The simple formula here is $(2^N/2) - 1$.

These concepts can perhaps be better understood with the example given in the following section and in Sect. 8.5.1.

Leibniz's dissertation on combinatorics, *De arte combinatoria* Leibniz (1666), was published in Leipzig nineteen years before J. S. Bach was born, and it includes the formula $2^n - 1$ for combinations given above, the formula $n!$ for permutations given in Sect. 5.1.2 as well as the mathematical techniques of partitioning (see Chap. 5). In 1636 Schwenter described a combination lock (Schwenter, Harsdörffer (1692) Vol. 1).

---

[3] The first combination might be the first 12 and another combination is the last 12, which is the opposite of the first.

8.2 Simple Proportions and Terminology

8.2.4 Solutions, Targets, Opposites and Complements

Now let us take a concrete example to define some more terms.

We shall use a contrived example with small lengths for simplicity, a set of six pieces named U, V, W, X, Y and Z with their lengths in bars as follows:

The total number of bars of all the pieces, the total length, is $5 + 10 + 15 + 8 + 7 + 15 = 60$.

There are 2^6-1 or 63 combinations in total.

We are interested in those pairs of combinations that are in a certain proportion such as 1:1 or 1:2.

For a proportion of $p{:}q$ these will be given by the two combinations of pieces hose lengths add up to:

Total Length × $\frac{p}{p+q}$	e.g. for 1:2	$60 \times \frac{1}{1+2} = 20$
and		
Total Length × $\frac{q}{p+q}$		$60 \times \frac{2}{1+2} = 40$

I call such a combination of pieces a **solution** (a solution to one of the equations above) and the two combinations that form the proportion a **solution pair**. A solution pair must contain all the pieces of the set with each piece only occurring in one solution of the pair. For the example in Table 8.2 a proportion of 1:2 has the following 3 solution pairs which divide the 60 bars into a 1:2 proportion or 20:40 bars (Table 8.3):

Table 8.2 A simple set of pieces with numbers of bars

Pieces:	U	V	W	X	Y	Z	Total length
Lengths:	5	10	15	8	7	15	60

Table 8.3 Example solution pairs for 1:2 proportion

Solution index	U	V	W	X	Y	Z	Sum	Proportion
	5	10	15	8	7	15	60	
1S	5		15				20	1
1C		10		8	7	15	40	2
2S	5			8	7		20	1
2C		10	15			15	40	2
3S	5					15	20	1
3C		10	15	8	7		40	2

The other solution of the pair always consists of the pieces not included in the first. For example, having found that U and W (5 and 15) add up to 20 we know that the others X, Y and Z will form the other solution with 10, 8, 7 and 15 adding up to 40 (rows 1S and 1C in Table 8.3). We say that the second solution is the **complement** of the first, as it contains those pieces that are not included in the first (we shall see the full reason for this term in Sect. 8.2.6).

We give each solution pair an index (shown in the first column)—both solutions of the pair have the same index number suffixed with an "S" or "C" to denote the solution or its complement.

When searching for a solution pair, we only need to search for one of the solutions, i.e. those that add up to 20. I call this the **target** value

$$\text{Target Value} = \text{Total Length} \times \frac{p}{p+q}$$

If we search for the other side of the proportion

$$\text{Target Value} = \text{Total Length} \times \frac{q}{p+q}$$

for 1:2 this would be 2:1, those that add up to 40, and we obtain the solution pairs in the opposite order. I call these the **opposite** solution pairs.

This gives rise to a special case for the 1:1 proportion which we saw in Sect. 8.2.3: since $p = q$ the target values for the solution pair and the opposite solution pair are the same and they cannot be differentiated.

We shall see how we go about finding all the solutions in Sect. 8.5.1.

Note that we generally show the combinations in rows rather than the columns used by Tatlow (2015) for greater convenience in later processing with a spreadsheet program such as Microsoft Excel, because:

(1) Excel is limited to 16,384 columns but can have over 1 million rows, and as we shall see, we need more rows than columns.
(2) Excel filters can be useful for further exploration, and these only work on columns.

It is easy to transpose the rows and columns in Excel to obtain a column format if required.

8.2.5 Symmetries, Signatures and Patterns

When examining the possible combinations of pieces from a set we might expect the composer to achieve the 1:1 or other proportion in some pleasing or symmetrical way. Tatlow (2015) devotes a whole chapter to "Symmetry, proportion and parallels".

8.2 Simple Proportions and Terminology

Having found all the possible combinations, we would therefore like to find those that have some kind of symmetry or other interesting pattern. There are various ways in which we can use the computer to assist with this.

8.2.6 Binary Signatures

To obtain a compact representation of each solution and thus reduce the amount of data to be processed we only need to know whether a piece falls in the upper or lower row of the solution pair because we have the list of lengths in the title row (see Table 8.3). This can be represented by a binary "Yes" or "No" or "1" or "0".

Applying this to Example 1 in Table 8.3 gives Table 8.4. The original length entries for the solutions can be obtained by multiplying the length in the title row by the 1 or 0 of the corresponding solution entry.

This is not only compact, but also lends itself to computer processing to detect patterns.

Writing the first solution pair as binary numbers gives:

```
101000
010111
```

I call these the binary **signatures** of the solutions. (We could go even further and convert this binary number to decimal as 40 and 23).

The lower row of the binary signature of each solution pair in Table 8.4 is simply the one's complement[4] of the upper row, i.e. all the bits inverted or flipped (0 changed to 1 and 1 changed to 0). This is another reason why I called it the "complement" in Sect. 8.2.4. We use this fact again later (in Sect. 8.5.3).

Table 8.4 Example binary signatures

Solution index	U	V	W	X	Y	Z
	5	10	15	8	7	15
1S	1	0	1	0	0	0
1C	0	1	0	1	1	1
2S	1	0	0	1	1	0
2C	0	1	1	0	0	1
3S	1	0	0	0	0	1
3C	0	1	1	1	1	0

---

[4]Since we do not use the two's complement, I simply use "complement" to refer to the one's complement.

Table 8.5 Examples of interesting patterns

Pattern	Signature	Complement
Palindrome (mirror image):	1 0 0 0 0 1	0 1 1 1 1 0

> Taking the complement of a binary number is a basic operation in computer hardware for performing binary arithmetic and is therefore comparatively fast. There are subtle differences between the one's complement and the two's complement.

Having a list of the binary signatures of a set of pieces, we can use it to find symmetries or other interesting **patterns**. For example, the third solution pair in Table 8.4 is symmetrical as shown in Table 8.5—the right half is the reverse of the left half, or it has reflection symmetry about the centre, or it is palindromic.

I use the word **template** to denote a binary sequence used to match against signatures to find patterns—more on this is Sect. 8.5.3.

8.3 Layers of Proportion

Section 8.2.4 showed how combinations of pieces can give a certain proportion such as 1:2 or 1:1. We mentioned in Sect. 8.1 that additional layers of proportion can be obtained by further dividing the solutions and their complements in the same proportion. This is illustrated in Fig. 8.7. The 2088 bars of book 1 of the Well Tempered Clavier are divided into a solution 1S and its complement 1C, each of 1044 bars, a 1:1 proportion. I call this Layer 1. These in turn are divided again, and we add a further layer of numbering to the index so that 1S is divided into a solution 1S_1S and its complement 1S_1C, each of 522 bars; 1C is likewise divided into 1C_1S and 1C_1C. This is Layer 2. We have now divided the 2088 bars into four equal sets of 522 bars each. We can do this again, giving eight sets of 261 bars

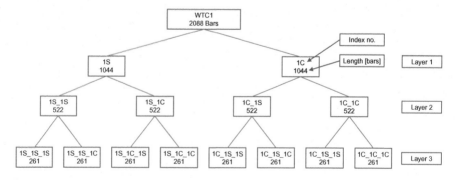

Fig. 8.7 Layers of 1:1 proportion

8.3 Layers of Proportion

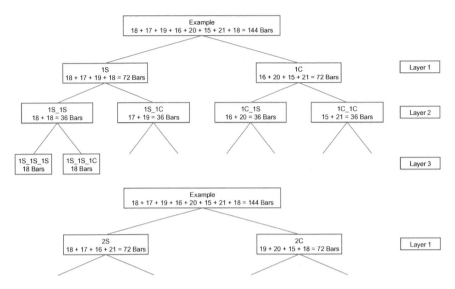

Fig. 8.8 Simple example of layer structure

each in Layer 3, numbered 1S_1S_1S, etc. These cannot be further divided in a 1:1 proportion because 261 is not an even number.

We could consider dividing lower layers in different proportions, for example the 261 bars of Layer 3 above could be divided in a 1:2 proportion with 87:174. However, we do not pursue this as it has no historical basis. We only consider layers in the same proportion.

Depending on the lengths of the individual pieces, the layers may not always be complete. This is illustrated with a simple example in Fig. 8.8 using eight pieces of lengths 18, 17, 19, 16, 20, 15, 21 and 18 bars. Solution Pair 1 is complete down to Layer 2, but for Layer 3 only 1S_1S can be further divided in a 1:1 proportion. Remember that there will normally be multiple solutions for a proportion, and another Solution Pair 2S and 2C in the lower part of the diagram cannot be further divided at all.

To be aesthetically pleasing, a layered proportion should be complete through all the desired layers, and I call this a **strict** solution. For the example in Fig. 8.8, solution pair 1 is strict for layer 2 but although there are solutions at layer 3, it is not strict for layer 3. To put it another way, a strict solution at a given layer includes all the pieces at that layer. If only one layer is being considered, all solutions are strict by nature. Figure 8.7 is an example that is strict over three layers.

Counting Solutions in Layers

Every solution has a complement and we count solution pairs, so a solution and its complement, e.g. 1S and 1C, are counted as one. This applies through all layers.

Since a solution and its complement may have different numbers of solutions at the next layer, the counts for the layers below layer 1 are not particularly meaningful and we cannot apply the combinatorial thinking of Sect. 8.2.3 to the lower layers because we do not know in advance how many first-layer combinations actually have solutions.

Pattern Matching in Layers

The pattern matching is generally applicable to all layers. We shall see how this works in more detail when we present the computer program.

8.4 Summary of Terms

Some terms were already defined in Sect. 1.3. The following definitions are ordered as they build upon each other rather than alphabetically.

Piece	A self-contained composition which has a defined length.
Set	A number of pieces for which the proportions are to be analysed. The set of pieces may be the works in a collection, the movements of a work, all the movements of the works in a collection, or any other arbitrary pieces brought together for analysis.
Length	The length of a piece, usually in bars.
Total Length	The total lengths of all the pieces in a set.
Combination	A subset of the pieces in a set.
Proportion	An integer relationship between the summed lengths of two combinations that partition the entire set, e.g. 1:2.
Proportional Count	An integer relationship between the numbers of pieces in two combinations that partition the entire set, e.g. 1:2.
Double Proportion	Having a proportion in the lengths and the same proportion count in the numbers of pieces that make up those lengths.
Target	The fraction of the total length of a set given by a proportion, e.g. for 1:2 the target is one-third of the total length.
Solution Pair	A pair of combinations which taken together include all the pieces in the set, and whose lengths are in the desired proportion to each other. The pair consists of the solution and its complement.
Solution	A combination of pieces, the length of which equals the target.
Complement	The combination containing those pieces of the set which are not in the solution.
Opposite	For a given proportion p:q the opposite solution pair is the pair having the q:p or opposite proportion.
Layer	A subdivision of a solution with the same proportion.

8.4 Summary of Terms

Strict Layer	A layer which divides all the solutions and complements of the layer above and so includes all the pieces of the original set.
Strict Proportion Count	Having Strict Layers of Proportion with all layers also having the Double Proportion.
Signature	A binary representation of a solution (or complement), denoting which pieces are in the solution (or complement) by 1s and those that are not in the solution (or complement) by 0s.
Pattern	A combination of particular interest, e.g. where the signature has a symmetry.
Template	A binary sequence used to match a complete or partial signature against patterns.

8.5 The Proportional Parallelism Explorer Program

Finding all the solutions for a given proportion is not manageable for any set of more than about 6 pieces for which one would have to try 63 combinations (or 31 for 1:1). A computer program was therefore developed to do this. Apart from searching for all possible solutions over multiple layers, the program can look for double proportions and patterns and perform Monte Carlo simulations. (It does not currently do triple or more proportions such as 1:1:1).

The complete user manual for the program is in Appendix B.

The program was initially developed in Microsoft Excel Visual Basic with the intention of making it readily available to users, but the complexity of the processing soon exceeded Excel's capabilities. The program was rewritten in Java, a language for developing programs to run on different systems such as Microsoft Windows, Apple Macintosh or UNIX derivatives such as LINUX. It was also extended to read the lengths of pieces from an input file, include a user interface where different parameters for the exploration of proportions can be set, selecting any proportion p:q and choosing the contents of the output files. The output was designed to be suitable for further processing in programs such as Excel, as this has useful filtering and sorting capabilities, or R, a powerful statistics program [see (Internet5)].

8.5.1 Solution Search

The underlying algorithm for the solution search was based on a source from the Internet (Internet30) which interestingly was for a completely different purpose, intended for processing payments coming into a company where a single payment could be for multiple invoices. The program was intended to find all possible combinations of invoices that would match the received payment. Here I adapted the algorithm to find all possible combinations of lengths of pieces that match a proportion of the total length of the overall set.

Having read in the titles and lengths of a set of pieces from an input file, the program calculates the sum of the lengths and the target for the desired proportion. The algorithm starts with the length of the first piece and then adds the lengths of subsequent pieces, trying each against the target until the target is reached. If the target is exceeded, the program stops that sequence and goes on to the next. (For the interested programmer, it uses recursion to call itself for each iteration).

Before starting, the program checks that the desired proportion is possible with the given set of pieces, and if not, stops with an error message. For example, if the sum of the lengths is an odd number, a 1:1 proportion cannot be obtained. The total length must be divisible by the sum of the two parts of the proportion, e.g. for 1:2 the total length must be divisible by 3.

To help understand how the program works, the above sample set of 6 simple pieces (Table 8.2) is shown in Table 8.6.

The sequence in which the combinations are tried is shown in the left column labelled "C". The sums of the lengths of the pieces in the combination are shown in the "Sum" column. Those that give a 1:2 proportion (target = 20) are combinations with combination numbers 18, 27 and 32. Those that give 1:1 (target = 30) are combinations with sequence numbers 3, 12 and 17.

For proportions other than 1:1, this will only find the first solution of each pair, and the program optionally delivers the complement to make the complete solution pair; e.g. for combination 18 in Table 8.6, the 1:2 proportion, the program would output the solution.

5		15				

and its complement

	10			8	7	15

To obtain the pairs in the opposite order, the program must be run with the opposite proportion, i.e. 2:1 instead of 1:2. For the given example of a total of 60 bars, this means searching for combinations that add up to the target of 40 (for 2:1) rather than 20 (for 1:2). Running the opposite proportion q:p with q > p will take longer due to the larger target—this will become clear below.

For the special case of the 1:1 proportion we make use of a consequence of the way the program works. Since the program works from left to right while adding up the pieces towards the target, the complements will be found in reverse order, i.e. from right to left starting at the bottom of the results. So, when a solution is the complement of its predecessor, we can stop the search as we are then finding the opposite solutions. In Table 8.6 combination 51 is the complement of combination

8.5 The Proportional Parallelism Explorer Program

Table 8.6 Demonstration with 6 simple lengths

C	U 5	V 10	W 15	X 8	Y 7	Z 15	Sum 60	Signature	C	U 5	V 10	W 15	X 8	Y 7	Z 15	Sum 60	Signature
1	5						5	100000	33		10					10	010000
2	5	10					15	110000	34		10	15				25	011000
3	5	10	15				30	111000	35		10	15	8			33	011100
4	5	10	15	8			38	111100	36		10	15	8	7		40	011110
5	5	10	15	8	7		45	111110	37		10	15	8	7	15	55	011111
6	5	10	15	8	7	15	60	111111	38		10	15	8		15	48	011101
7	5	10	15	8		15	53	111101	39		10	15		7		32	011010
8	5	10	15		7		37	111010	40		10	15		7	15	47	011011
9	5	10	15		7	15	52	111011	41		10	15			15	40	011001
10	5	10	15			15	45	111001	42		10		8			18	010100
11	5	10		8			23	110100	43		10		8	7		25	010110
12	5	10		8	7		30	110110	44		10		8	7	15	40	010111
13	5	10		8	7	15	45	110111	45		10		8		15	33	010101
14	5	10		8		15	38	110101	46		10			7		17	010010
15	5	10			7		22	110010	47		10			7	15	32	010011
16	5	10			7	15	37	110011	48		10				15	25	010001
17	5	10				15	30	110001	49			15				15	001000
18	5		15				20	101000	50			15	8			23	001100
19	5		15	8			28	101100	51			15	8	7		30	001110
20	5		15	8	7		35	101110	52			15	8	7	15	45	001111
21	5		15	8	7	15	50	101111	53			15	8		15	38	001101
22	5		15	8		15	43	101101	54			15		7		22	001010
23	5		15		7		27	101010	55			15		7	15	37	001011
24	5		15		7	15	42	101011	56			15			15	30	001001
25	5		15			15	35	101001	57				8			8	000100
26	5			8			13	100100	58				8	7		15	000110
27	5			8	7		20	100110	59				8	7	15	30	000111
28	5			8	7	15	35	100111	60				8		15	23	000101
29	5			8		15	28	100101	61					7		7	000010
30	5				7		12	100010	62					7	15	22	000011
31	5				7	15	27	100011	63						15	15	000001
32	5					15	20	100001									

17, so the program stops there. Using this method, the opposite solutions cannot be obtained by reversing the proportion, as reversing 1:1 is still 1:1, so a special setting is provided to cater for this, i.e. not to stop after finding the first complement. This will yield combinations 51, 56 and 59 in addition to 3, 12 and 17.

The program is additionally optimised so that when the target value is exceeded, it does not try any further combinations starting with the same pieces. This is the reason why running with 2:1 will take longer than 1:2—it will skip less combinations.

Let us follow the example in Table 8.6 for a 1:1 proportion (Target = 30) using the combination numbers in the left "C". The program will try combination 1 (5), then 2 (5 + 10 = 15), then 3 (5 + 10 + 15 = 30). This is a solution, so the program outputs this and optionally its complement and performs the pattern matching (see Sect. 8.5.3). Since adding further pieces would exceed the target, the program skips the combinations that start with $5 + 10 + 15$ and continues with combination 11. In the end, it only tries 38 of the 63 possible combinations. (In fact, we should exclude combination number 6 containing all the pieces to give 62 possible combinations, but adding code to do that would not make the program significantly faster and would add unnecessary complexity).

Further efficiency could be gained by sorting the pieces in descending order of length. This ensures that the target is reached or exceeded with the minimum number of tries so that the program takes the least possible time to complete. However, the solutions must be returned to their original order before displaying them, calculating the signature and looking for any patterns. It was found that the time taken to reverse the sorting gives diminishing returns with increasing number of pieces in the set, and the advantage is not worthwhile, so this has not been implemented. For sets where the order of the pieces is not important, the user may sort the pieces by descending length to optimise the processing time. However, if the order is important and patterns are being sought based on the order, the original order must be used.

Figure 8.9 shows the actual output of the program for the above test data for the 1:2 proportion after import to Excel. The only formatting that has been done is to centre some of the columns. The index column gives each solution pair an index number for reference in the pattern matching (see below)—it is not the same as the combination number above as not all combinations are solutions. The Layer column gives the layer and will be explained later, the Sum column gives the sum of the lengths, the Count column gives the number of pieces in the solution and the PropCount column shows whether the number of pieces is in the same proportion

Index	Layer	Sum	Count	PropCount	U	V	W	X	Y	Z
1S	1	20	2	Y	5		15			
1C	1	40	4	Y		10		8	7	15
2S	1	20	3		5			8	7	
2C	1	40	3			10	15			15
3S	1	20	2	Y	5					15
3C	1	40	4	Y		10	15	8	7	

Program Version: 6.0c
Input File: /Users/alan/OneDrive/PropParResults/Z_Book/6test_for book.txt.
Total Length: 60. Proportion: 1:2. Target: 20. Tried: 32.
Include Complements: Yes. 1:1 Opposites: No.
Layers: 1. Strict Layers: Yes. Only First Strict: No.
PropCount: Yes. Strict PropCount: No.
Strict Proportion Solutions: 3.
Strict PropCount Solutions: 2.
Solutions Layer 1: 3.
Patterns: 1.
Processing Time: 0,004 Seconds without Matrix.

Fig. 8.9 Output of solution search

as the lengths (double proportion). The output optionally includes a summary at the end showing the parameters used for the run.

8.5.2 Solutions Search Through Layers

To find additional layers of proportion when a solution is found, for example solution 3 in Table 8.6 for a 1:1 proportion, we take the lengths of the solution, here 5, 10 and 15 as three pieces with a total length of 30 bars and feed these back into the routine with a new target of 15, set according to the proportion of the total length of these pieces. We also do this for the complement of the solution, 8, 7 and 15. The results are expanded back into their original columns. The first solution is shown as a diagram in Fig. 8.10 as a diagram and the complete solution as the program output in Fig. 8.11 (there are three 1:1 solutions and each can be broken down into the second layer in both the solution and complement branches, so they are all strict). For more layers this is repeated recursively for each solution and complement found at every layer.

For larger works using two or more layers the number of solutions rapidly becomes very large and therefore takes a long time to process and produces large output files, often too large to open completely in Excel. The strict solutions are the most elegant, so the program has options for only outputting the strict solutions and for stopping when one strict solution has been found. This also reduces the amount of data produced.

8.5.3 Pattern Matching

When it comes to finding patterns in the data, although the human eye and brain is one of the best pattern recognition tools there is, the sheer volume of data and the monotony of its appearance makes this very tedious if not impossible.

The program uses the binary signatures (see Sect. 8.2.5) to find interesting patterns. The user can define binary templates that are compared to the signatures of the

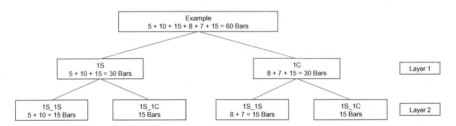

Fig. 8.10 Diagram for example of 2 layers

Index	Layer	Sum	Count	PropCount	U	V	W	X	Y	Z
1S	1	30	3	Y	5	10	15			
1C	1	30	3	Y				8	7	15
1S_1S	2	15	2		5	10				
1S_1C	2	15	1				15			
1C_1S	2	15	2					8	7	
1C_1C	2	15	1							15
2S	1	30	4		5	10		8	7	
2C	1	30	2				15			15
2S_1S	2	15	2	Y	5	10				
2S_1C	2	15	2	Y				8	7	
2C_1S	2	15	1	Y			15			
2C_1C	2	15	1	Y						15
3S	1	30	3	Y	5	10				15
3C	1	30	3	Y			15	8	7	
3S_1S	2	15	2		5	10				
3S_1C	2	15	1							15
3C_1S	2	15	1				15			
3C_1C	2	15	2					8	7	
Program Version: 6.0c										
Input File: /Users/alan/OneDrive/PropParResults/Z_Book/6test_for book.txt.										
Total Length: 60. Proportion: 1:1. Target: 30. Tried: 38.										
Include Complements: Yes. 1:1 Opposites: No.										
Layers: 2. Strict Layers: No. Only First Strict: No.										
PropCount: Yes. Strict PropCount: No.										
Strict Proportion Solutions: 3.										
Strict PropCount Solutions: 0.										
Solutions Layer 1: 3. Layer 2: 6.										
Patterns: 1.										
Processing Time: 0,002 Seconds without Matrix.										

Fig. 8.11 Output of solution search 1:1 with 2 layers

solutions. These can either be compared exactly, or the template pattern can be sought in any position in the signature (shifted).

"**Exact**" matching only gives a match if the template exactly matches the signature of the solution. The number of bits given in the template must be the same as the number of pieces. For example, the pattern of the first 1:2 solution shown in Tables 8.4 and 8.6 and Fig. 8.9 would be found with the following template:

```
101000
```

"**Shift**" matching will give a match if a shorter template occurs at any position in the solution's signature. The template is compared with the signature from the left, and then shifted one position to the right and compared again, and so on until the template is on the right.

8.5 The Proportional Parallelism Explorer Program

> Shifting a binary number left or right is a common operation in computer hardware and is therefore comparatively fast. It is equivalent to multiplying or dividing the binary number by two, just as shifting a decimal number left or right is the equivalent of multiplying or dividing it by 10.

For example,
the template

`101`

compared with the signature

`010100010100`

will compare as follows as it is shifted to the right:

```
010100010100 – no match
010100010100 – match
010100010100 – no match
010100010100 – no match
010100010100 – no match
010100010100 – no match
010100010100 – no match
010100010100 – match
010100010100 – no match
010100010100 – no match
```

The template must have fewer bits than the number of pieces. (If it had the same number, an exact match would be more appropriate).

The program also checks the template against the complement of the signature. This ensures that all potential patterns are found without having to run the opposite proportion separately.

Some patterns cannot easily be defined with the above method so some "built in" pattern matching functions are available.

The following are currently implemented:

- **Left=Right** - Left half = Right half, i.e. the right half of the signature is the same as the left half, e.g. for six pieces `110 110`, or in mathematical terms, translation symmetry.
- Left half is the reverse of the right half, i.e. the right half of the signature is the same as the left half, but in reverse order or mirror image, e.g. for six pieces `110 011` - a palindrome or reflection symmetry. I call this pattern "**Left=thgiR**".

Index	Layer	Pattern	PropCount	A	B	C	D	E	F	G	H
1S	1	Four >3^	Y	18	17	19					18
1S_1S	2	Left=thgiR+^	Y	18							18
1C_1S	2	Left=thgiR+^	Y				16	20			
4S	1	Left=Right+^	Y	18			16	20			18
4S	1	Left=thgiR+^	Y	18			16	20			18
4S_1S	2	Left=thgiR+^	Y	18							18
5S	1	Four >1^	Y	18					15	21	18
5S_1S	2	Left=thgiR+^	Y	18							18

Fig. 8.12 Output of pattern matching

- **Count** to show whether the solution contains a given number of pieces.

The means of specifying this pattern matching for the program input is given in the user manual Sects. B.3.4–B.3.7.

Additionally the program will flag solutions with a **Proportional Count** if the number of pieces in the solution is equivalent to the chosen proportion of the number of pieces in the set, i.e. that the proportion m:n is achieved with m pieces against n pieces as well as in the lengths (double proportion). For example the first 1:1 solution pair in Table 8.6, combination number 3, will match this requirement as the 1:1 (30:30) length proportion is achieved with 1:1 (3:3) pieces. This is particularly important as it is a further layer of proportional parallelism in addition to that of the numbers of bars, and the program has an option for restricting the search to strict solutions which are also strict with respect to the proportional count. A proportional count is indicated with an additional column in the solutions output—see Fig. 8.11.

Figure 8.12 shows the pattern matching output in Excel for a different example. Solutions 1S matches a shift template named "Four" consisting of four 1's shifted right 3 places (>3) in the complement (^).

The PropCount column shows whether there is an additional proportion in the number of pieces, so solution 1S is shown as having an additional 1:1 proportion in the 4:4 pieces. This is in the solutions output (above) and is repeated in the patterns output as it is useful for filtering in Excel.

The exact meanings of the symbols are explained in the user manual (Appendix B).

8.5.4 Pattern Matching in Layers

The pattern matching is applied to the layers and will be shown with the appropriate index for the layer, e.g. 1S_1S, and the layer number. In Fig. 8.12, solution 1S_1S in layer 2 matches the mirror image "Left=thgiR" template in both the solution and its complement (denoted by + ^).

For layer 2 the PropCount column shows the additional proportion in the count of 2:2 pieces (equally dividing the 4 pieces of solution 1S).

8.5 The Proportional Parallelism Explorer Program

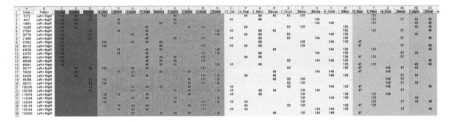

Fig. 8.13 Example colour coding of Mass in B minor

Note that although the solution search in the lower layers is performed with the reduced number of pieces from the previous layer, the pattern matching acts with the pieces in their original positions in the overall sequence of pieces. This gives consistent results over all layers.

8.5.5 Colour Coding for Visual Pattern Recognition

Another method which may facilitate the visual recognition of patterns is to colour code the columns in some way. Figure 8.13 shows an example of the above patterns with each main section of the Mass in B minor, Kyrie, Gloria, Credo, etc. given a different colour in Excel.

This can be combined with the use of filters in Excel and may be useful in spotting patterns involving the sections of the composition. Due to the large amount of data I have not found this to be very effective.

8.5.6 Monte Carlo Simulation

The program can also be used to explore random samples of sets of pieces with Monte Carlo simulations (see Sect. 7.5).

The composer was not restricted to any particular lengths for the pieces in a collection or work, and it is interesting to compare the proportions found in the real collections or works with random sets of pieces, as we shall see in Chap. 9.

To do this we select the number of pieces in the set, the minimum and maximum number of bars in each piece and the number of samples we wish to produce. The program will generate sample sets of pieces with random numbers of bars, and for each sample find the solutions as described in Sect. 8.5.1. In this mode the program only counts the solutions, it does not output them, and it does not search for patterns. The order of the pieces is therefore not important, and the lengths of the pieces in the samples are sorted into descending order to optimise the solution search. It outputs a table showing for each of the sample sets the random lengths of the pieces, and for each layer the number of solutions and the number of solutions that also have the proportion in the number of pieces.

The random lengths generated will produce some sets for which the given proportion is impossible, e.g. a set with a total length which is an odd number cannot have a 1:1 proportion. The program ignores these samples and generates the desired number of samples which can give the proportion. The number ignored is shown and can be taken into account if desired.

This permits the following type of scenario to be explored: the Well Tempered Clavier Book 1 has 24 works with lengths between 53 and 159 bars. How many solutions would a random collection of 24 pieces with lengths between 53 and 159 bars have on average (or maybe 50 and 170, as the composer was not restricted to a particular minimum or maximum number of bars)? We can use the program to generate say 100,000 random sample collections (sets) and then use Excel or R to plot the histogram and perform further analysis.

To facilitate histogram plotting, the program can also output the data for a histogram of the number of samples for each number of solutions and data for a histogram of the lengths generated. The lengths histogram is used to validate that the random lengths are similar to a uniform distribution—if it is not sufficiently uniform, more samples are required as we saw in Fig. 7.4, Sect. 7.5. These can be used to easily plot graphs in Excel or R.

An example of the program output for our simple example above over two layers is shown in Fig. 8.14 (only the beginning and the end of the 100,000 rows are shown).

The six pieces are given titles R1–R6 (for Random) and their lengths are shown in columns A–F. Each row is a sample set of six pieces with random lengths between 3 and 15 bars, to correspond with our simple example in Sect. 8.5.1.

Column G shows the number of solutions that give the 1:1 proportion in the lengths and column H shows the number of solutions that also have the proportion in the numbers of pieces for each random sample set in layer 1. Columns I and J show these for layer 2. Column K shows the time in seconds taken to process each sample (which is 0—as shown at the end, the program took less than two seconds to process all 100,000 samples). The overall summary is shown at the end with the main statistical values minimum, maximum, mean and standard deviation for each layer. Other parameters such as the percentile in which a real value falls can be calculated from the data with Excel functions.

The results histogram output is shown in Fig. 8.15. For each number of solutions in column A, the number of samples that had that number of solutions is given in the columns B to E. For example, looking at row 8, in layer 1, 83 samples had 6 solutions with the 1:1 proportion in the numbers of bars and 88 samples had 6 solutions with the proportion in the numbers of pieces. In layer 2, 1208 samples had 6 solutions with the proportion in the numbers of bars and 55 samples had the proportion in the numbers of pieces. This is show graphically below the table.

We see that for this 1:1 example for layer 1, one solution occurs most frequently and the average (the mean given at the end of Fig. 8.14) is just over 1, which fits well with our real data set.

8.5 The Proportional Parallelism Explorer Program

	A	B	C	D	E	F	G	H	I	J	K
1	R1	R2	R3	R4	R5	R6	L1_Bars	L1_Pieces	L2_Bars	L2_Pieces	Seconds
2	15	14	10	8	6	3	1	1	1	0	0
3	14	11	8	4	4	3	2	1	2	0	0
4	13	11	7	3	3	3	1	0	0	0	0
5	15	13	12	8	4	4	1	0	0	0	0
6	15	14	11	10	9	5	0	0	0	0	0
7	10	7	6	5	4	4	1	1	0	0	0
8	15	15	14	13	10	9	2	2	0	0	0
9	14	12	12	9	6	3	0	0	0	0	0
10	14	10	9	7	5	3	2	1	1	1	0
11	14	13	13	10	3	3	0	0	0	0	0

	A	B	C	D	E	F	G	H	I	J	K
99988	15	14	10	7	6	4	1	1	1	0	0
99989	13	12	11	9	6	5	1	1	0	0	0
99990	11	9	8	7	6	3	1	1	1	0	0
99991	11	11	8	4	3	3	0	0	0	0	0
99992	10	10	8	7	7	6	2	2	0	0	0
99993	14	12	9	8	5	4	2	1	1	1	0
99994	13	11	11	9	7	5	0	0	0	0	0
99995	13	12	11	8	8	4	1	1	0	0	0
99996	12	10	8	6	5	3	1	0	1	1	0
99997	10	9	9	5	4	3	0	0	0	0	0
99998	13	13	10	8	5	3	3	2	6	2	0
99999	14	11	11	7	5	4	1	1	0	0	0
100000	11	8	6	6	5	4	1	1	0	0	0
100001	15	14	13	12	10	8	1	1	0	0	0
100002	Program Version: 6.0e
100003	Tried: 100 000. Ignored: 300 554.
100004	Pieces: 6. Samples: 100 000.
100005	Minimum Length: 3. Maximum Length: 15.
100006	Proportion: 1:1. Strict Layers: No. Strict PropCount: No. Layers: 2.
100007	Layer 1:
100008	Min. Solutions: 0. Max. Solutions: 7.
100009	Mean: 1,05005. Std. Deviation: 0,96921.
100010	Min. PropCounts: 0. Max. PropCounts: 6.
100011	Mean PropCounts: 0,84725. Std. Deviation: 0,91473.
100012	Layer 2:
100013	Min. Solutions: 0. Max. Solutions: 16.
100014	Mean: 0,40753. Std. Deviation: 0,98431.
100015	Min. PropCounts: 0. Max. PropCounts: 15.
100016	Mean PropCounts: 0,11787. Std. Deviation: 0,43708.
100017	Processing Time: 1,462 Seconds.
100018	

Fig. 8.14 Main output of Monte Carlo simulation

The program also outputs a lengths histogram file with the lengths of all the pieces and the frequencies with which each length occurred. This is shown in Fig. 8.16. The lengths are in the left column and the frequency with which each length occurred is in the right column. This is plotted as a histogram beside the data.

The histogram plots must be created and formatted by the user. The diagrams shown were produced with R.

	A	B	C	D	E
1	Solutions	L1_Bars	L1_Pieces	L2_Bars	L2_Pieces
2	0	33604	43946	74456	91530
3	1	36532	32443	17497	5776
4	2	22904	19921	5988	2253
5	3	5534	2496	647	380
6	4	1160	1106	26	5
7	5	178	0	0	0
8	6	83	88	1208	55
9	7	5	0	0	0
10	8	0	0	0	0
11	9	0	0	0	0
12	10	0	0	0	0
13	11	0	0	172	0
14	12	0	0	0	0
15	13	0	0	0	0
16	14	0	0	0	0
17	15	0	0	1	1
18	16	0	0	5	0
19					

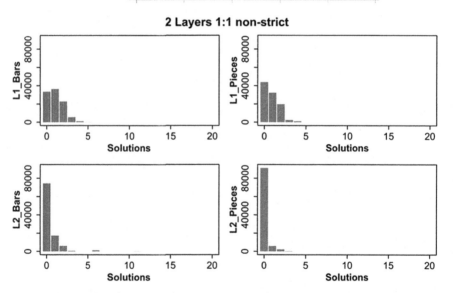

Fig. 8.15 Results histogram output of Monte Carlo simulation

8.5 The Proportional Parallelism Explorer Program

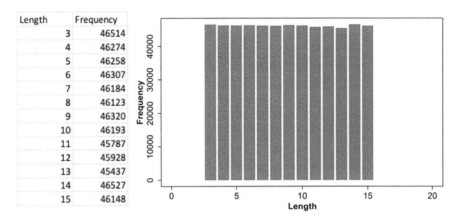

Length	Frequency
3	46514
4	46274
5	46258
6	46307
7	46184
8	46123
9	46320
10	46193
11	45787
12	45928
13	45437
14	46527
15	46148

Fig. 8.16 Lengths histogram output of Monte Carlo simulation

Monte Carlo Simulation with Strict Layers.

The Monte Carlo simulation acts on the desired number of layers. In non-strict mode the results and results histogram output files have columns for each layer. In strict mode there are only columns for the first layer, as we are only interested in the number of strict solutions and not the number of possibilities at the lower layers.

Chapter 9
Applying the Methods to the Well Tempered Clavier Book 1 BWV 846–869

9.1 Preamble

To show how these techniques can be applied, we shall follow in Ruth Tatlow's footsteps and see what additional light can be shed on the results in Tatlow (2015).

The detailed statistical analysis of the works and collections is not given to easy reading, so these are placed in Appendix A. We will use the Well Tempered Clavier Book 1 (WTC1) as an example to show how the methods can be applied in practice and then summarise the results overall. WTC1 was chosen because the set of 24 pieces is sufficiently large to give some interesting data, but not too large to run the required simulations, and there is no doubt as to how to count the bars—there is only one repeat and no da capos or time signature changes, and if the repeat is counted there can be no 1:1 or 1:2 proportion.

9.2 Solutions

To start with, we take each prelude and fugue together as a piece. Tatlow (2015) presents one way of making the bars in a subset of the 24 preludes and fugues of WTC1 add up to 1044 (half of the total of 2088 bars for the 1:1 proportion)—the first shaded columns in Fig. 9.1. The program finds 14,191 different ways. That is, using the terminology defined in Sect. 8.2.4, there are 14,191 1:1 solution pairs. The figure of 14,191 does not include the opposite pairs obtained by swapping the two columns. We will come to the other 1:1 columns later when we consider layers.

The first five solution pairs found by the program are shown in Table 9.1 which is laid out to correspond with Fig. 9.1 for easier comparison. The fifth example has a parallel proportion in the number of pieces (proportional count).

Table 6.1 Structure of *WTC I*. Autograph score. P 415

BWV	Key	Title	Bars	Title	Bars	Total	1:1	1:1	1:1	2:1	B-A-C-H
846	C major	Praeludium 1	35	Fuga 1	27	62	62		62	131	131
847	C minor	Praeludium 2	38	Fuga 2	31	69	69	69			
848	C♯ major	Praeludium 3	104	Fuga 3	55	159	159	159		313	
849	C♯ minor	Praeludium 4	39	Fuga 4	115	154	154		154		
850	D major	Praeludium 5	35	Fuga 5	27	62	62		62	132	
851	D minor	Praeludium 6	26	Fuga 6	44	70		70		70	
852	E♭ major	Praeludium 7	70	Fuga 7	37	107		107		107	234
853	E♭ minor	Praeludium 8	40	Fuga 8	87	127		127		127	
854	E major	Praeludium 9	24	Fuga 9	29	53		53		53	136	
855	E minor	Praeludium 10	41	Fuga 10	42	83		83		83		
856	F major	Praeludium 11	18	Fuga 11	72	90		90		90	170	
857	F minor	Praeludium 12	22	Fuga 12	58	80	80	80			
858	F♯ major	Praeludium 13	30	Fuga 13	35	65		65		65	129	
859	F♯ minor	Praeludium 14	24	Fuga 14	40	64		64	64		
860	G major	Praeludium 15	19	Fuga 15	86	105		105	105	158	
861	G minor	Praeludium 16	19	Fuga 16	34	53		53	53		
862	A♭ major	Praeludium 17	44	Fuga 17	35	79		79	79	149	
863	G♯ minor	Praeludium 18	29	Fuga 18	41	70		70		70	
864	A major	Praeludium 19	24	Fuga 19	54	78		78	78	193	193
865	A minor	Praeludium 20	28	Fuga 20	87	115	115	115			
866	B♭ major	Praeludium 21	20	Fuga 21	48	68	68		68	167	
867	B♭ minor	Praeludium 22	24	Fuga 22	75	99	99	99			
868	B major	Praeludium 23	19	Fuga 23	34	53			53	176	176
869	B minor	Praeludium 24	47	Fuga 24	76	123	123	123			
Totals			819		1269	2088	1044 : 1044	522 : 522	522 : 522	1392 : 696	500

Fig. 9.1 Proportions in WTC1. From Tatlow (2015) reproduced with permission of the Licensor through PLSclear

The numbers for some more proportions of the total 2088 bars are shown in Table 9.2. There are no 2:3 proportions because the total length of 2088 is not divisible by 5. The number of solution pairs decreases rapidly as the q in the 1:q proportion increases and the target decreases.

Some other proportions were explored briefly in Sect. 8.1.

We will now apply the methods described in Chap. 7.

9.3 Probability

At first sight, finding the large number of 14,191 1:1 solutions for WTC1 would appear to indicate that an arbitrary combination of lengths would give a 1:1 proportion by chance. However, this must be seen in relation to the overall number of possibilities, i.e. the total number of possible ways in which it is possible to split the collection into two groups that we showed from the combinations in Sect. 8.2.3—this is 8,388,607.

> The probability of one of our 14,191 1:1 proportions occurring by chance is therefore 14,191/8,388,607 i.e. 0.0017 or 0.17% or 1 in 591.

9.3 Probability   119

Table 9.1 First 5 examples of the 14,191 different 1:1 proportions of WTC1

	Index	Example 1		Example 2		Example 3		Example 4		Example 5	
		1S	1C	2S	2C	3S	3C	4S	4C	5S	5C
	Length	1044	1044	1044	1044	1044	1044	1044	1044	1044	1044
	Count	11	13	11	13	11	13	11	13	12	12
										Y	Y
Proportional count											
PF1	C major	62	62		62		62		62		62	
PF2	C minor	69	69		69		69		69		69	
PF3	C# major	159	159		159		159		159		159	
PF4	C# minor	154	154		154		154		154		154	
PF5	D major	62	62		62		62		62		62	
PF6	D minor	70	70		70		70		70		70	
PF7	E♭major	107	107		107		107		107		107	
PF8	E♭minor	127	127		127		127		127			127
PF9	E major	53		53		53		53		53	53	
PF10	E minor	83		83		83		83		83	83	
PF11	F major	90	90		90			90		90	90	
PF12	F minor	80	80			80		80		80		80
PF13	F# major	65		65	65		65		65		65	
PF14	F# minor	64	64			64	64			64		64

(continued)

Table 9.1 (continued)

	Index		Example 1		Example 2		Example 3		Example 4		Example 5	
			1S	1C	2S	2C	3S	3C	4S	4C	5S	5C
	Length		1044	1044	1044	1044	1044	1044	1044	1044	1044	1044
	Count		11	13	11	13	11	13	11	13	12	12
PF15	G major	105		105		105	105			105		105
PF16	G minor	53		53		53		53		53		53
PF17	A♭major	79		79	79			79		79		79
PF18	G# minor	70		70		70		70			70	
PF19	A major	78		78		78		78		78		78
PF20	A minor	115		115		115		115		115		115
PF21	B♭major	68		68		68		68		68		68
PF22	B♭minor	99		99		99		99	99			99
PF23	B major	53		53		53		53		53		53
PF24	B minor	123		123		123		123		123		123

9.3 Probability

Table 9.2 Proportions of bars for WTC1

Proportion	Target	No. of solution pairs
1:1	1044	14,191
1:2	696	9041
1:3	522	2106
1:5	348	196
1:7	261	29
1:8	232	24
1:11	174	4

(If we consider finding either a solution or its complement, then we must consider this in relation to the complete set of combinations, i.e. $28,382/(2^{24} - 1) = 28,382/16,777,215$, which is the same.).

This is visualised in Fig. 9.2 which on the left shows the numbers as squares of corresponding areas—imagine the number as a quantity of dots distributed evenly over the square. The large light grey square represents all the 8,388,607 possible combinations, and the small dark grey square represents the 14,191 solutions. Alternatively, the graphic on the right shows the relationship as a dartboard the area of whole dartboard (outer yellow circle) represents the 8,388,607 combinations, the red inner disc is the bullseye and the yellow circle around it is the area representing the 14,191 solutions. The probability of finding a 1:1 proportion by chance from WTC1 is about the same as hitting the bullseye when throwing a dart at a dartboard without aiming (ignoring the darts that miss the board altogether).

If we consider the 1:2 proportion, there were 9041 solution pairs, and these must be considered in relation to the full set of 16,777,215 possibilities, i.e. a probability of 0.05%.

Fig. 9.2 Infographics of WTC1 1:1 proportions within all combinations. Dartboard image by Vectorportal.com with annotations by author

Layers

Looking back at Fig. 9.1, Tatlow (2015) gives two columns for a second layer of 1:1 solutions with 522 bars each. The relevant part is shown in Fig. 9.3, the last two columns.

This is shown as program output in Fig. 9.4. The first 1:1 column with 1044:1044 is solution 415S and 415C, the next column is 415S_2S and 415S_2C and the last column is 415C_1S and 415C_1C. There are other alternatives for the second layer of solution 415—three for 415S and seven for 415C.

Table 6.1 Structure of *WTC I*. Autograph score. P 415

BWV	Key	Title	Bars	Title	Bars	Total	1:1		1:1		1:1	
846	C major	Praeludium 1	35	Fuga 1	27	62	62			62		
847	C minor	Praeludium 2	38	Fuga 2	31	69	69		69			
848	C♯ major	Praeludium 3	104	Fuga 3	55	159	159		159			
849	C♯ minor	Praeludium 4	39	Fuga 4	115	154	154			154		
850	D major	Praeludium 5	35	Fuga 5	27	62	62			62		
851	D minor	Praeludium 6	26	Fuga 6	44	70		70			70	
852	E♭ major	Praeludium 7	70	Fuga 7	37	107		107			107	
853	E♭ minor	Praeludium 8	40	Fuga 8	87	127		127			127	
854	E major	Praeludium 9	24	Fuga 9	29	53		53		53		
855	E minor	Praeludium 10	41	Fuga 10	42	83		83			83	
856	F major	Praeludium 11	18	Fuga 11	72	90		90		90		
857	F minor	Praeludium 12	22	Fuga 12	58	80	80		80			
858	F♯ major	Praeludium 13	30	Fuga 13	35	65		65			65	
859	F♯ minor	Praeludium 14	24	Fuga 14	40	64		64		64		
860	G major	Praeludium 15	19	Fuga 15	86	105		105		105		
861	G minor	Praeludium 16	19	Fuga 16	34	53		53		53		
862	A♭ major	Praeludium 17	44	Fuga 17	35	79		79		79		
863	G♯ minor	Praeludium 18	29	Fuga 18	41	70		70			70	
864	A major	Praeludium 19	24	Fuga 19	54	78		78		78		
865	A minor	Praeludium 20	28	Fuga 20	87	115	115		115			
866	B♭ major	Praeludium 21	20	Fuga 21	48	68		68		68		
867	B♭ minor	Praeludium 22	24	Fuga 22	75	99		99		99		
868	B major	Praeludium 23	19	Fuga 23	34	53		53			53	
869	B minor	Praeludium 24	47	Fuga 24	76	123		123			123	
Totals			819		1269	2088	1044 : 1044		522 : 522		522 : 522	

Fig. 9.3 Second Layer in WTC1 from Tatlow (2015). Reproduced with permission of the Licensor through PLSclear

Index	Layer	Sum	Count	PropCount	01_Cm	02_cm	03_C♯m	04_c♯m	05_Dm	06_dm	07_E♭m	08_d♯m	09_Em	10_em	11_Fm	12_fm	13_F♯m	14_f♯m	15_Gm	16_gm	17_A♭	18_g♯m	19_Am	20_am	21_B♭m	22_b♭m	23_Bm	24_bmi
415S	1	1044	11		62	69	159	154	62							80								115	68	99	53	123
415C	1	1044	13							70	107	127	53	83	90		65	64	105	53	79	70	78			99	53	
415S_1S	2	522	6		62	69	159									80												
415S_1C	2	522	5					154	62	70	107	127	53	83	90		65	64	105	53	79	70	78	115	68			123
415S_2S	2	522	6		62			154	62																68		53	123
415S_2C	2	522	5			69	159									80								115		99		
415S_3S	2	522	5		62			154																115	68			123
415S_3C	2	522	6			69	159		62							80										99	53	
415C_1S	2	522	6							70	107	127		83			65					70						
415C_1C	2	522	7										53		90			64	105	53	79		78					
415C_2S	2	522	7							70	107		53		90					53	79	70						
415C_2C	2	522	6									127		83			65	64	105				78					
415C_3S	2	522	7							70	107		53				65			53	79	70	78					
415C_3C	2	522	6									127		83	90			64	105	53								
415C_4S	2	522	7								107		53	83	90		65				79							
415C_4C	2	522	6							70		127						64	105	53		70	78					
415C_5S	2	522	6							70		127	53	83					105		79		78					
415C_5C	2	522	7								107				90		65	64		53		70						
415C_6S	2	522	7							70	107						65			53	79	70	78					
415C_6C	2	522	6									127	53	83	90			64	105									
415C_7S	2	522	7							70		127	53	83	90		65	64		53								
415C_7C	2	522	6								107								105		79	70	78					

Fig. 9.4 Second layer in WTC1 1:1 as program output

Index	Layer	Sum	Count	'opCou#1	Cm_d2_cmi3	C#m.4_c#mi5	Dm_d6_dmi7	Ebm.8_d#m!9	Em≥0_emi.1	Fm≤2_fmi3	F#m.4_f#mi5	Gm≤.6_gmi7	Abm.8_g#m 9	Am≥0_ami1	Bbm2_bbm 3	Bm≤4_bmir
6S	1	1044	12	Y	62	69	159	154	62	70	107		53	83		79		78		68		
6C	1	1044	12	Y							127				90	80	65	64	105	53		70		115		99	53	123
6S_2S	2	522	6	Y	62	69	159			70			83			79				
6S_2C	2	522	6	Y				154	62		107	53					78		68		
6S_3S	2	522	6	Y	62	69		154	62		107						78		68		
6S_3C	2	522	6	Y			159			70		53	83			79	78			
6S_6S	2	522	6	Y	62		159		62		107	53				79				
6S_6C	2	522	6	Y		69		154		70			83				78		68		
6S_7S	2	522	6	Y	62			154			107	53					78		68		
6S_7C	2	522	6	Y		69	159		62	70			83			79				
6C_2S	2	522	6	Y							127				90		65	64		53							123
6C_2C	2	522	6	Y												80			105			70		115		99	53	
6C_3S	2	522	6	Y							127				90		65	64									53	123
6C_3C	2	522	6	Y												80			105	53		70		115		99		

Fig. 9.5 The first 1:1 two-layer solution with parallel number of pieces

The program finds 12,688 strict solutions for two layers, which is 89% of the 14,191 single-layer solutions. The probability of finding a strict two-layer solution is therefore

$$\frac{14{,}191}{8{,}388{,}607} \times \frac{12{,}688}{14{,}191} = 0.00151 \text{ or } 0.151\%$$

There will be more to say on this in Sect. 9.6.

For the 1:2 proportion over two strict layers, 4910 strict solutions are found.

Another form of layers of proportion is with the number of pieces. For a single layer, 6825 of the 14,191 1:1 solutions consist of 12:12 pieces, that is 48% or nearly half of them. For two layers 5783 solutions have the 1:1 proportion in the number of pieces as well as in the lengths over both layers, with multiple alternatives at layer 2, shown in Fig. 9.5. This is 5783/12,688 = 46% of the strict two-layer solutions.

For the 1:2 proportion 1918 of the single layer solutions also have the 1:2 proportion in the number of pieces, 1918/4910 = 39%.

9.4 Monte Carlo Simulation

The above considers the actual bar lengths of WTC1. Let us also consider other possibilities, after all, Bach could have chosen any lengths. Based on the fact that the shortest three pieces in WTC1 have 53 bars and the longest piece has 159 bars, we would like to perform a Monte Carlo simulation (see Sect. 7.5) by generating a random sample of say, 100,000 collections of 24 pieces with lengths between 53 and 159 bars. As Bach was not restricted to these minimum and maximum lengths, we could arbitrarily extend the range, e.g. to between 50 and 160 or between 30 and 200 bars, but since we cannot rationally decide on a suitable range, we will stay within the given parameters throughout for the sake of consistency. The program gives the following results.

The distribution of the individual lengths of the 240,000 pieces (100,000 samples of 24 pieces each) is shown as a histogram in Fig. 9.6 and this is close enough to a uniform distribution for us to be convinced that the randomisation is adequate and we have a sufficiently large number of samples.

The distribution of the numbers of single-layer solutions is shown in Fig. 9.7. Note

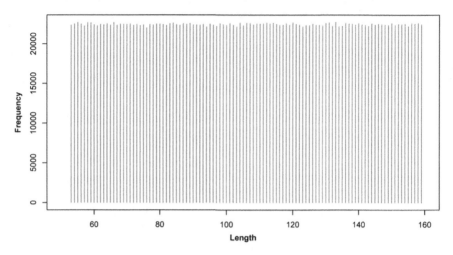

Fig. 9.6 Distribution of individual lengths

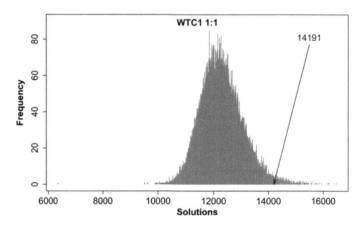

Fig. 9.7 Monte Carlo simulation of WTC1 1:1

that although the individual random lengths were generated to be equally likely, the overall picture is similar to a normal distribution[1] which is as expected (see Sect. 7.5).

For the 100,000 random collections we have the following results:

Minimum number of Solutions 6338
Average number of Solutions 12,266
Maximum number of Solutions 16,490

There are no random collections which have no solutions, and with repeated runs of the program none have been observed so far.

---

[1] This is explained by the Central Limit Theorem.

9.4 Monte Carlo Simulation

Thus it can be claimed that in virtually any random set of 24 pieces like WTC1: There will be many ways of dividing them up into a 1:1 proportion.
The probability of finding a 1:1 proportion by chance in any such random collection is:
on average $12{,}266/8{,}388{,}607 = 0.146\%$,
at least $6338/8{,}388{,}607 = 0.076\%$ and
at most $16{,}490/8{,}388{,}607 = 0.2\%$.

So how does the real WTC1 compare with this?

The real WTC1 has 14,191 1:1 solutions (see Table 9.2) and this is in the 99th percentile of the random results (shown in Fig. 9.7), i.e. over 99% of random collections have less solutions than the real WTC1 or less than 1% of random collections have more solutions than the real WTC1.

Percentile

A percentile is the value below which a given percentage of the samples fall, e.g. 99% of the values fall below the 99th percentile. Note that this is 99% of the area under the curve, not 99% of the way along the x-axis.

Some conclusions are:
We should not be surprised that proportions can be found in WTC1, because proportions can be found in any random collection with equivalent lengths. The probability of finding a proportion in the real WTC1 is much higher than on average for equivalent random collections.

Repeating the exercise for the 1:2 proportion we have

Minimum number of Solutions: 4543
Average number of Solutions: 7536
Maximum number of Solutions: 10,152

Thus it can be claimed that virtually any random set of 24 pieces like WTC1:
- There will be many ways of dividing them up into a 1:2 proportion.
- The probability of finding a 1:2 proportion by chance in any such random collection is:

on average $7536/16{,}777{,}215 = 0.045\%$,
at least $4543/16{,}777{,}215 = 0.027\%$ and

at most $10{,}152/16{,}777{,}215 = 0.06\%$.

The number of 9041 solutions for the real WTC1 is again well above the average in the 97th percentile (Fig. 9.8).

These particular 100,000 sample collections did not include the real lengths of WTC1. This is not surprising, considering that the total population of all combinations of the range of lengths is 5×10^{48} as we saw in Sect. 7.5.

Layers

We can run the Monte Carlo simulation over two layers, but this naturally takes a long time, so we do not perform 100,000 simulations.

The Monte Carlo simulation (1000 samples) for 1:1 with two strict layers gives

Minimum number of Solutions:	5054
Average number of Solutions:	10,989
Maximum number of Solutions:	14,894

With 10,000 samples:

Minimum number of Solutions:	1073
Average number of Solutions:	10,992
Maximum number of Solutions:	14,987

For two layers of 1:2 using 10,000 samples the figures are (ran 4 h):

Minimum number of Solutions:	0
Average number of Solutions:	3287
Maximum number of Solutions:	6643

For the Preludes alone (7 h):

Minimum number of Solutions: 3196

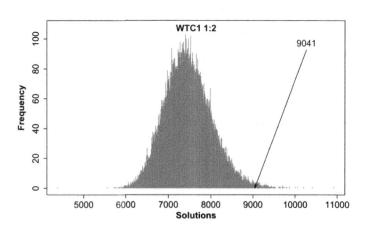

Fig. 9.8 Distribution of 1:2 Solutions for 24 Pieces

9.4 Monte Carlo Simulation

Average number of Solutions: 9377
Maximum number of Solutions: 17,635

And for the Fugues alone (6 h):

Minimum number of Solutions: 2148
Average number of Solutions: 7039
Maximum number of Solutions: 13,696

These are shown graphically with the real results in Fig. 9.9.

We could ask: if we believe that Bach intentionally created a 1:1 or other proportion, which of the thousands of solutions did he find? These techniques cannot provide the answer. The historical approach followed by Tatlow (2015), which analyses the manuscripts and copies to trace the alterations to the numbers of bars made by Bach over time and uses manual techniques to find the proportions, is more informative.

9.5 Hypothesis Testing and Significance

If we apply the thinking on hypotheses testing and significance from Sect. 7.3 to the 1:1 proportions in WTC1, the null hypothesis would be: "The 1:1 proportion is due to chance". Classical hypothesis testing is difficult to apply here, because although we have a Monte Carlo simulation for which we can obtain the mean and standard deviation of the number of solutions in our random sample sets, we do not have an

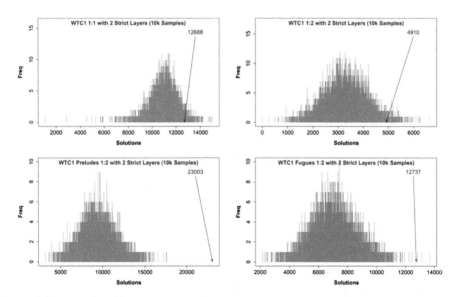

Fig. 9.9 WTC1 Monte Carlo simulations of 2 Layers

Fig. 9.10 Simple Bayesian network for 1:1 proportion in WTC1

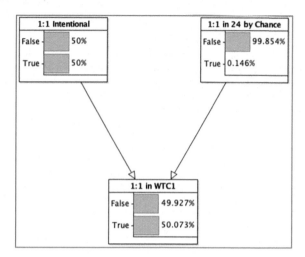

equivalent sample set for the real collection with which to compare—there is only one. However, we can formulate the above results as follows:

- The probability of a collection such as WTC1 having proportions is close to 1. Almost any collection of 24 pieces with the same range of lengths will have proportions.
- The probability of a specific combination of the pieces in the real WTC1 producing a single-layer proportion by chance is 0.17% for 1:1 and 0.05% for 1:2. For strict two-layer proportions the figures are 0.15% for 1:1 and 0.03% for 1:2.
- This probability of finding a proportion in the real collection is slightly higher than that obtained for random collections: for 1:1 it is 0.17% instead of 0.146% and for 1:2 it is 0.05% instead of 0.045%. For two-layer strict proportions we have 0.13% for 1:1 and 0.02% for 1:2.

Is the fact that WTC1 has proportions due to chance? Yes, because any equivalent random collection has proportions.

If a specific proportional combination is found, can we reject the null hypothesis that this is due to chance? Yes, these are well below the threshold of 5% commonly used for sociology or medical trials, so we could reject the null hypothesis on this basis and say it is not due to chance. On the other hand, it is far above the threshold used by particle physicists, and they would hold back before announcing this as a significant result.

9.6 Bayes Theorem

A first Bayesian network (which we introduced in Sect. 7.6) for 1:1 proportions in WTC1 created with AgenaRisk 10, is shown in Fig. 9.10. This shows our assump-

9.6 Bayes Theorem

tion that the probability of finding a 1:1 proportion in WTC1 (lower box) depends on the probability of 1:1 being intentional (upper left box) and the probability of a 1:1 proportion in 24 random pieces occurring by chance (upper right box), the dependencies being shown by the arrows.

- The node "1:1 Intentional" is True if a 1:1 proportion was created intentionally and False if not. We assign this a prior assumption of 50% for each case, i.e. we do not know whether Bach intentionally created a 1:1 proportion.
- The node "1:1 in 24 by Chance" is True if a 1:1 proportion is obtained by chance and False if not. We assign to these the probabilities we know from our Monte Carlo simulation (section above), i.e. True = 0.146%. False = 100 − 0.146 = 99.854% (the latter is automatically assigned by the program).
- The node "1:1 in WTC1" is True if there is a 1:1 proportion in WTC1 and False if not. This node depends on the other two as shown by the arrows, i.e. the probability of finding a 1:1 proportion in WTC1 depends on the probability of one being intentionally created and on the probability of one occurring by chance. For this node we must fill in the Node Probability Table for each combination, 8 values rather than just the two each for the other nodes. This is shown in Fig. 9.11.

To fill this in we consider whether WTC1 will have 1:1 proportions given the states of the other two nodes. The top left entry (A) says: if "1:1 Intentional" is False AND "1:1 in 24 by Chance" is False, then "1:1 in WTC1 is certainly False", i.e. a probability of 1.0. The three lower right entries (B) say: if a 1:1 occurs either by chance or by intention, then WTC1 certainly will have a 1:1 (True with a probability of 1.0). The overall probabilities (the columns) must add up to 1.0.

This initial state in Fig. 9.10 already reflects the combined probabilities of "1:1 Intentional" and "1:1 in 24 by Chance" in that the result "1:1 in WTC1" is slightly more likely to be true.

Now we can use the power of Bayesian inference to test a scenario. Since we actually observe that we do have 1:1 proportions in WTC1 we can enter this observation in the "1:1 in WTC1" node as shown in Fig. 9.12.

The program will then recalculate the probabilities using this new information as shown in Fig. 9.13. This shows a slightly increased probability that the observation

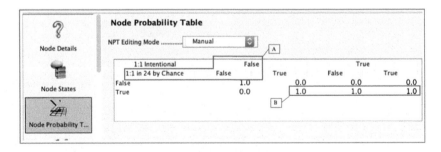

Fig. 9.11 Node probability table for "1:1 in WTC1"

is due to chance (from 0.146 to 0.292%) and a large increase in the probability that it was intentional—from 50 to 99.854%.

If we test the fictitious converse observation, i.e. that we found no 1:1 proportions in WTC1, the result, as shown in Fig. 9.14, is that there were certainly no intentional 1:1 and no chance of finding 1:1, which is as we would expect.

We can also take a more sceptical approach and set our prior assumption to a 5% chance that the 1:1 proportions are intentional—Fig. 9.15 shows this, with a correspondingly high initial probability that there will be no 1:1s in WTC1.

If we then enter the observation as before (Fig. 9.16), we again see that the probability that it was intentional is greatly increased to over 97%, although not quite as high as when we had no prior opinion on the intentionality. The probability of it being by chance increases to 2.84%.

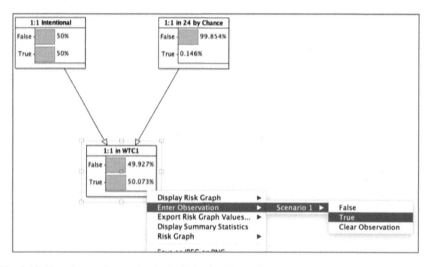

Fig. 9.12 Entering an observation in the Bayesian Network

9.6 Bayes Theorem

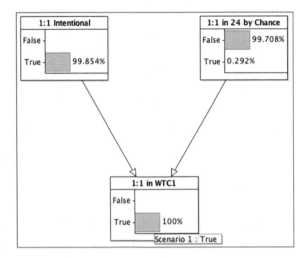

Fig. 9.13 Bayesian network with result of observation of 1:1 proportion in WTC1

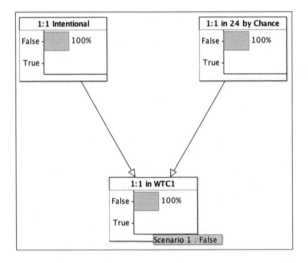

Fig. 9.14 Bayesian network testing the case of not finding 1:1 in WTC1

This simple model does not really add anything new to our intuitive conclusion that if a 1:1 proportion is unlikely to occur by chance it is probably intentional, but supports it with mathematical logic and enables us to experiment with different prior assumptions.

The Bayesian network can of course be extended with any other dependencies.

Layers

You probably noticed that the equation for the probability of finding a two-layer solution contained a potential cancellation of terms leaving the number of two-layer

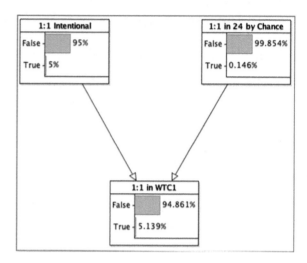

Fig. 9.15 Simple Bayesian network for WTC1 with sceptical prior

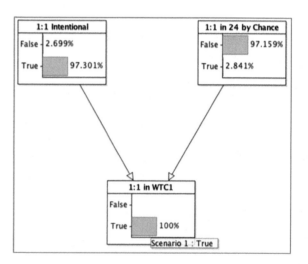

Fig. 9.16 Simple Bayesian network for WTC1 sceptical with observation

solutions divided by the number of layer 1 combinations. We can apply Bayesian thinking to confirm this. We are talking about the probability of finding a layer 2 solution P(L2) given that we have found a layer 1 solution with probability P(L1). Using Bayes' theorem from Sect. 7.6

$$P(L2|L1) = \frac{P(L1|L2) \times P(L2)}{P(L1)}$$

But P(L1|L2), the probability of finding a layer 1 solution given that we have a layer 2 solution, is 1—there cannot be a layer 2 solution without a layer 1 solution from which it is derived. Therefore

$$P(L2|L1) = \frac{P(L2)}{P(L1)}$$

For WTC1 1:1 this works out as

$$P(L1) = \frac{14{,}191}{8{,}388{,}607} = 0.00169 \text{ and } P(L2) = \frac{12{,}688}{8{,}388{,}607} = 0.00151$$

$$P(L2|L1) = \frac{0.00151}{0.00169} = 0.89 \text{ or } 89\%$$

And from the second equation above.
$P(L2) = P(L2|L1) \times P(L1) = 0.89 \times 0.00169 = 0.00151$ as before.

This confirms that, even though we do not know the number of possible combinations at layer 2, we can calculate the probability of finding layer 2 solutions using the combinations of layer 1:

$$P(L2) = \frac{\text{No. of } L2 \text{ solutions}}{\text{No. of } L1 \text{ combinations}}$$

9.7 Patterns

For the Well Tempered Clavier Book 1 (WTC1) Tatlow (2015) shows an example of a 1:1 proportion where the first 5, the twelfth and the last 5 give the same number of bars as the rest, the 6 before the twelfth and the 7 after it (see the first shaded column in Fig. 9.1). This is almost symmetrical but not quite, but a different symmetry is shown in conjunction with the Inventions and Sinfonias—see Sect. 10.13.

If Bach intentionally divided them into a 1:1 or 1:2 proportion, we could ask: would he not have used some logical and elegant division as a basis for dividing up the pieces? Would he not have applied the numeric proportion to some aspect of the overall design? E.g. shown in Table 9.3:

- the first 12 and the second 12,
- the major and minor keys (these are divided into 1:1 by their very nature but not in parallel with the lengths),
- the keys of the diatonic scale and others,
- dividing between the preludes and fugues (which are again by nature in a 1:1 proportion), which is already shown in Tatlow's table as being 819:1269, and with the greatest common divisor of 9 gives an unsatisfactory 91:141 proportion.

or, particularly relevant given the purpose of the collection:

Table 9.3 Proportions of combinations in WTC1 if intentional

No	Key	No. of bars	First half	Second half	Major keys	Minor keys	Diatonic keys	Other keys	Mean tone keys	Other keys	Older keys Sachs	Other keys	Inv and Sin keys	Other keys
PF1	C major	62	62		62		62		62		62		62	
PF2	C minor	69	69			69	69			69	69		69	
PF3	C# major	159	159		159			159		159		159		159
PF4	C# minor	154	154			154		154		154		154		154
PF5	D major	62	62		62		62		62		62		62	
PF6	D minor	70	70			70	70		70		70		70	
PF7	E♭ major	107	107		107			107		107		107	107	
PF8	E♭ minor	127	127			127		127		127		127		127
PF9	E major	53	53		53		53			53		53	53	
PF10	E minor	83	83			83	83			83	83		83	
PF11	F major	90	90		90		90		90				90	
PF12	F minor	80	80			80	80			80		80	80	
PF13	F# major	65		65	65			65		65		65		65
PF14	F# minor	64		64		64		64		64		64		64
PF15	G major	105		105	105		105		105		105		105	
PF16	G minor	53		53		53	53		53		53		53	
PF17	A♭ major	79		79	79			79		79		79		79
PF18	G# minor	70		70		70		70		70		70		70

(continued)

9.7 Patterns

Table 9.3 (continued)

No	Key	No. of bars	First half	Second half	Major keys	Minor keys	Diatonic keys	Other keys	Mean tone keys	Other keys	Older keys Sachs	Other keys	Inv and Sin keys	Other keys
PF19	A major	78		78	78		78			78	78		78	
PF20	A minor	115		115		115	115		115		115		115	
PF21	B♭major	68		68	68			68	68		68		68	
PF22	B♭minor	99		99		99		99		99	99			99
PF23	B major	53		53	53		53			53		53		53
PF24	B minor	123		123		123	123			123	123		123	
Totals		2088	1116	972	981	1107	1096	992	625	1463	1077	1011	1218	870
Greatest common divisor			36	27	9		8		1		3		174	
Proportion			31	27	109	123	137	124	625	1463	359	337	7	5

- the keys playable in mean tone tuning versus those only playable in a well-tempered tuning. I could not find a definitive list of playable mean tone keys. They sound more and more dissonant until you get to the remotest keys with the intolerable "wolf" dissonance, and it is a matter of taste how far away from the tuned home key you modulate. I started with C D d F G g a B[2] and then found Sachs who gives some lists of keys from earlier collections such as C c D d F G g A a B h, or these with the addition of E♭. Wolff (1991) in his essay on "The Clavier-Übung Series" mentions the "still-prevailing system of unequal temperament, that is, containing generally no more than three sharps or flats", i.e. C c D d E♭e F f# G g A a B h. None of these were found when sought by the explorer program as a pattern.
- The keys used in the Inventions and Sinfonias, regarding these as an easier collection for Wilhelm Friedemann Bach before graduating to the Well Tempered Clavier. This gives a 5:7 proportion, the best yet, but not particularly meaningful.

None of these give an interesting proportion.

The input file templates of the explorer program for these and other patterns are shown in Table 9.4.

We mentioned some other potentially interesting proportions in Sect. 8.1.

Of course, it may be, that if there is any intentionality, it may simply be that of creating a proportion without any further meaning. It is up to the musicologist to answer these questions—my intention is simply to point out the possibilities.

The built-in patterns for symmetries give the following results for the 1:1 proportion.

The same sequence in both halves (Left = Right) occurs 4 times shown in Table 9.5 (plus 4 opposites not shown). These all have a parallel proportion in the number of pieces (Count). The Index shows the number of the solution and complement within the solutions.

The pattern of the second half being the mirror image of the first half (Left = thgiR) is shown in Table 9.6 and occurs in 5 combinations (plus 5 opposites not shown). These all also have the parallel proportion in the number of pieces (Count).

In Fig. 9.2, where the probabilities were visualised, the 5 possibilities would be represented as a square so small as to be invisible.

For a 1:2 proportion there are two solutions that give the mirror image pattern shown in Table 9.7. Again, these have the parallel proportion in the number of pieces.

Looking at the second layer of proportions, there are symmetrical patterns there, but none of the strict solutions has a symmetry in both layers.

Similarly to the proportions themselves, we could also ask: if Bach intentionally created a pattern, which of the possibilities did he intend? (Table 9.8).

---

[2]Using the German notation where B is the English B♭ and H is the English B. Upper case is major and lower case is minor.

9.7 Patterns   137

Table 9.4 Pattern matching input file for WTC1

```
// Pattern Templates
// Bit Positions 10s        0000000001111111111222222
// Bit Positions 1s         1234567890123456789012334
Template  Bach      Exact   0100000000000000000011001   // Bmaj, amin, cmin, hmin
Template  bach      Exact   0100000000000000000010101   // bmin, amin, cmin, hmin
Template  BACH      Exact   1000000000000000000101010   // Bmaj, Amaj, Cmaj, Hmaj
Template  2138      Shift   1101001101111111            // bach sequences of 1s in numeric alphabet
Template  Tatlow    Exact   1111100000100000001111      // 1:1 given in Tatlow p.161
Template  12InARow  Shift   111111111111                // 12 pieces in a row
Template  MajMin    Exact   101010101010101010101010    // Major vs. Minor keys
Template  Diatonic  Exact   110010011100110010011       // Diatonic vs. Chromatic keys
Template  MeanTone  Exact   100010000100001100011000    // Mean Tone vs. Well Tempered playable keys
Template  Sachs1    Exact   110011000110001100111001    // From Sachs, 1984 C c D d e F G g A a B h
Template  Sachs2    Exact   110011001100011001111001    // From Sachs, 1984 C c C D d Es e F G g A a B h
Template  InvSin    Exact   110011101110011001111001    // Inventions & Sinfonias / WFB-Büchlein
BuiltIn   Left=Right                                    // Left half = Right half
BuiltIn   Left=thgiR                                    // Left half = Right half reversed (mirror image)
```

Table 9.5 Combinations of WTC1 1:1 with same sequence in both halves

			Pattern 1		Pattern 2		Pattern 3		Pattern 4	
Index			7731S	7731C	8379S	8379C	11742S	11742C	11889S	11889C
Length			1044	1044	1044	1044	1044	1044	1044	1044
Count			12	12	12	12	12	12	12	12
PF1	C maj	62	62		62		62		62	
PF2	C min	69		69		69		69		69
PF3	C# maj	159	159		159			159		159
PF4	C# min	154	154		154		154		154	
PF5	D maj	62		62		62		62		62
PF6	D min	70	70			70	70		70	
PF7	E♭ maj	107	107			107	107		107	
PF8	E♭ min	127		127		127	127			127
PF9	E maj	53	53		53			53		53
PF10	E min	83		83	83			83	83	
PF11	F maj	90		90	90		90			90
PF12	F min	80		80		80		80	80	
PF13	F# maj	65	65		65		65		65	
PF14	F# min	64		64		64		64		64
PF15	G maj	105	105		105			105		105
PF16	G min	53	53		53		53		53	
PF17	A♭ maj	79		79		79		79		79
PF18	G# min	70	70			70	70		70	
PF19	A maj	78	78			78	78		78	
PF20	A min	115		115		115	115			115
PF21	B♭ maj	68	68		68			68		68
PF22	B♭ min	99		99	99			99	99	
PF23	B maj	53		53	53		53			53
PF24	B min	123		123		123		123	123	

9.7 Patterns

Table 9.6 Combinations of WTC1 1:1 with mirror image

			Pattern 1		Pattern 2		Pattern 3		Pattern 4		Pattern 5	
Index			1472S	1472C	2258S	2258C	10844S	10844C	11698S	11698C	13314S	13314C
Length			1044	1044	1044	1044	1044	1044	1044	1044	1044	1044
Count			12	12	12	12	12	12	12	12	12	12
PF1	C maj	62	62		62		62		62		62	
PF2	C min	69	69		69			69		69		69
PF3	C# maj	159	159		159			159		159		159
PF4	C# min	154		154		154	154		154			154
PF5	D maj	62	62				62	62			62	62
PF6	D min	70	70		70		70		70			70
PF7	E♭ maj	107		107	107			107	107		107	
PF8	E♭ min	127		127	127		127		127		127	
PF9	E maj	53		53	53		53	53		53			53
PF10	E min	83		83	83		83		83		83		83
PF11	F maj	90	90		90			90		90	90	
PF12	F min	80		80		80		80		80	80	
PF13	F# maj	65		65		65		65		65	65	
PF14	F# min	64	64		64			64		64	64	
PF15	G maj	105		105		105		105		105		105
PF16	G min	53		53		53	53		53			53
PF17	A♭ maj	79		79		79	79		79		79	
PF18	G# min	70		70	70			70	70		70	
PF19	A maj	78	78		78		78		78			78
PF20	A min	115	115				115	115		115	115	
PF21	B♭ maj	68		68		68	68		68			68
PF22	B♭ min	99	99		99			99		99		99
PF23	B maj	53	53		53			53		53		53
PF24	B min	123	123		123		123		123		123	

Table 9.7 Combinations of WTC1 1:2 with mirror image

			Pattern 1		Pattern 2	
	Index		7798S	7798C	8349S	8349C
	Length		696	1392	696	1392
	Count		8	16	8	16
PF1	C major	62		62		62
PF2	C minor	69		69		69
PF3	C# major	159		159		159
PF4	C# minor	154		154		154
PF5	D major	62	62			62
PF6	D minor	70		70	70	
PF7	E♭ major	107	107			107
PF8	E♭ minor	127		127	127	
PF9	E major	53		53		53
PF10	E minor	83	83		83	
PF11	F major	90	90		90	
PF12	F minor	80		80		80
PF13	F# major	65		65		65
PF14	F# minor	64	64		64	
PF15	G major	105	105		105	
PF16	G minor	53		53		53
PF17	A♭ major	79		79	79	
PF18	G# minor	70	70			70
PF19	A major	78		78	78	
PF20	A minor	115	115			115
PF21	B♭ major	68		68		68
PF22	B♭ minor	99		99		99
PF23	B major	53		53		53
PF24	B minor	123		123		123

Table 9.8 Summary of solutions and patterns for WTC1

	WTC1 one layer			WTC1 two layers (Strict)		
	Solutions	Left = Right + Left = thgiR	Proportion count	Solutions	Left = Right + Left = thgiR	Proportion count
1:1	14,191	4 + 5	6825	12,688	0 + 0	5783
1:2	9041	0 + 2	3896	4910	0 + 1	0

9.8 Preludes and Fugues Separately

We could also consider all the preludes and fugues separately giving 48 pieces, but the time needed to run this makes it prohibitive. (We will find proportions another way later.) The 24 preludes and the 24 fugues can also be considered as two sets of pieces. Neither can have 1:1 solutions as they both have an odd number of bars, but there are many 1:2 solutions—see Tables 9.9, 9.10 and 9.11.

Fugues

Kramer notes that there is a 1:2 proportion within the fugues that puts the first 8 against the second 16 giving a double proportion. The program shows that this is the first of 5058 solutions that have 8 against 16 pieces (i.e. 1:2), and of these, 8 have a symmetry see Table 9.10. Kramer's pattern and the 8 symmetrical solutions with the additional parallel proportion in the number of pieces is shown in Table 9.12.

Taking these to a second layer of 1:2 proportions, there are 12,737 strict solutions for the fugues, and 4 of these have a Left = Right symmetry in both layers.

102—Left = Right and PropCount in layer 1,
7031—Left = Right and PropCount in layer 1,
8214—Left = Right and PropCount in layer 1,
12,011—Left = Right and PropCount in layer 1.

The first is shown in Table 9.15. Columns 102S and 102C are the layer 1 solution and its complement which divide the lengths into 423 and 846 bars. Beside these are the second layer which divides the solution into 141 and 282 bars and the complement into 282 and 564 bars.

The 24 pieces cannot have a proportion in the number of pieces in the second layer, because after dividing the 24 into 8 and 16, this layer is no longer divisible by 3.

Preludes

Kramer does not mention the preludes in this connection, but for 1:2 there are also over 5000 1:2 solutions which also have 8 against 16 pieces. The first eight are shown in Table 9.13. Of the 15 symmetrical solutions, 7 also have the 1:2 proportion in the number of pieces—see Tables 9.14 and 9.15.

The second layer has 23,003 strict 2-layer solutions and plenty of individual symmetries and proportions in the numbers of pieces, but no symmetries or proportional numbers of pieces that are complete over both layers.

It would also be possible to take books 1 and 2 together, but for WTC2 there is no definitive source for the lengths.

Table 9.9 Proportion of WTC1 preludes and fugues separately

Proportion	No. of solution pairs Preludes	No. of solution pairs Fugues
1:1	–	–
1:2	10,273	16,144

Table 9.10 Proportions in WTC1 fugues

	WTC1 fugues one layer			WTC1 fugues two layers (Strict)		
	Solutions	Left = Right + Left = thgiR	Proportion count	Solutions	Left = Right + Left = thgiR	Proportion Count
1:1	–	–	–	–	–	–
1:2	16,144	5 + 3	5058	12,737	3367 + 953	0

Table 9.11 Proportions in WTC1 preludes

	WTC1 preludes one layer			WTC1 preludes two layers (Strict)		
	Solutions	Left = Right + Left = thgiR	Proportion count	Solutions	Left = Right + Left = thgiR	Proportion Count
1:1	–	–	–	–	–	–
1:2	26,864	7 + 8	5128	23,003	2089 + 1538	0

Table 9.12 Kramer's and the symmetrical 1:2 patterns in lengths and pieces for WTC1 fugues

		(Kramer)		Pattern 2		Pattern 3		Pattern 4		Pattern 5	
	Index	1S	1C	102S	102C	406S	406C	543S	543C	7031S	7031C
	Length	423	846	423	846	423	846	423	846	423	846
	Count	8	16	8	16	9	15	10	14	8	16
F1	C maj	27	27		27		27		27		27
F2	C min	31	31		31		31		31		31
F3	C# maj	55	55		55		55		55		55
F4	C# min	115	115		115				115		115
F5	D maj	27	27			27	27			27	27
F6	D min	44	44			44		44	44			44
F7	E♭ maj	37	37			37		37		37		37
F8	D# min	87	87			87		87		87		87
F9	E maj	29		29		29	29		29			29
F10	E min	42		42		42		42	42			42
F11	F maj	72		72		72		72	72			
F12	F min	58		58		58	58			58		58
F13	F# maj	35		35	35			35		35		35
F14	F# min	40		40	40			40	40		40	
F15	G maj	86		86	86			86		86	86	
F16	G min	34		34	34		34			34	34	
F17	A♭ maj	35		35		35		35	35		35	
F18	G# min	41		41		41		41		41		41
F19	A maj	54		54		54		54		54		54
F20	A min	87		87		87	87			87		87
F21	B♭ maj	48		48		48		48	48			48
F22	B♭ min	75		75		75	75			75		75
F23	B maj	34		34		34		34		34		34
F24	B min	76		76		76		76		76		76

(continued)

9.8 Preludes and Fugues Separately

Table 9.12 (continued)

	Index		Pattern 6		Pattern 7		Pattern 8		Pattern 9	
			8214S	8214C	12011S	12011C	12299S	12299C	13394S	13394C
	Length		423	846	423	846	423	846	423	846
	Count		8	16	8	16	8	16	8	16
F1	C maj	27		27		27		27		27
F2	C min	31	31			31		31		31
F3	C# maj	55	55		55		55			55
F4	C# min	115		115		115		115	115	
F5	D maj	27		27		27		27		27
F6	D min	44		44	44			44		44
F7	E♭ maj	37		37	37		37		37	
F8	D# min	87		87		87	87			87
F9	E maj	29	29			29		29	29	
F10	E min	42		42		42		42		42
F11	F maj	72		72	72			72	72	
F12	F min	58	58			58	58			58
F13	F# maj	35		35		35	35			35
F14	F# min	40	40			40		40		40
F15	G maj	86	86		86			86		86
F16	G min	34		34		34		34	34	
F17	A♭ maj	35		35		35	35			35
F18	G# min	41		41	41		41			41
F19	A maj	54		54	54			54	54	
F20	A min	87		87		87		87		87
F21	B♭ maj	48	48			48		48	48	
F22	B♭ min	75		75		75	75			75
F23	B maj	34		34	34			34	34	
F24	B min	76	76			76		76		76

Table 9.13 First 8 of 5128 1:2 patterns in lengths and pieces for WTC1 preludes

		Pattern 1		Pattern 2		Pattern 3		Pattern 4	
	Index	99S	99C	100S	100C	101S	101C	118S	118C
	Length	273	546	273	546	273	546	273	546
	Count	8	16	8	16	8	16	8	16
P1	C major	35	35		35		35		35
P2	C minor	38	38		38		38		38
P3	C# major	104		104		104		104	
P4	C# minor	39	39		39		39		39
P5	D major	35	35		35		35		35
P6	D minor	26	26		26		26		26
P7	E♭ major	70		70		70		70	
P8	E♭ minor	40	40		40		40		40
P9	E major	24		24		24		24	24
P10	E minor	41	41		41		41		41
P11	F major	18		18		18		18	
P12	F minor	22		22		22		22	
P13	F# major	30		30		30		30	
P14	F# minor	24		24		24		24	
P15	G major	19	19			19		19	
P16	G minor	19		19	19			19	
P17	A♭ major	44		44		44		44	
P18	G# minor	29		29		29		29	29
P19	A major	24		24		24		24	
P20	A minor	28		28		28		28	
P21	B♭ major	20		20		20		20	
P22	B♭ minor	24		24		24		24	
P23	B major	19		19		19	19		19
P24	B minor	47		47		47		47	47

(continued)

9.8 Preludes and Fugues Separately

Table 9.13 (continued)

			Pattern 5		Pattern 6		Pattern 7		Pattern 8	
	Index		123S	123C	149S	149C	156S	156C	158S	158C
	Length		273	546	273	546	273	546	273	546
	Count		8	16	8	16	8	16	8	16
P1	C major	35	35		35		35		35	
P2	C minor	38	38		38		38		38	
P3	C# major	104		104		104		104		104
P4	C# minor	39	39		39		39		39	
P5	D major	35	35		35		35		35	
P6	D minor	26	26		26		26		26	
P7	E♭ major	70		70		70		70		70
P8	E♭ minor	40		40		40		40		40
P9	E major	24		24		24		24		24
P10	E minor	41	41			41		41		41
P11	F major	18		18		18		18		18
P12	F minor	22		22		22		22		22
P13	F# major	30	30			30		30		30
P14	F# minor	24		24	24			24		24
P15	G major	19		19		19		19		19
P16	G minor	19		19		19		19		19
P17	A♭ major	44		44		44		44		44
P18	G# minor	29	29		29		29		29	
P19	A major	24		24		24	24			24
P20	A minor	28		28		28		28		28
P21	B♭ major	20		20		20		20		20
P22	B♭ minor	24		24		24		24	24	
P23	B major	19		19		19		19		19
P24	B minor	47		47	47			47	47	

Table 9.14 Symmetrical patterns in lengths and pieces with 1:2 in WTC1 preludes

	Index		Pattern 1		Pattern 2		Pattern 3		Pattern 4	
			2090S	2090C	8142S	8142C	16592S	16592C	17328S	17328C
	Length		273	546	273	546	273	546	273	546
	Count		8	16	8	16	8	16	8	16
P1	C major	35	35		35			35		35
P2	C minor	38		38		38		38		38
P3	C# major	104		104		104	104		104	
P4	C# minor	39		39		39	39			39
P5	D major	35		35		35		35		35
P6	D minor	26	26		26		26		26	
P7	E♭ major	70		70	70			70		70
P8	E♭ minor	40	40			40		40		40
P9	E major	24		24		24		24	24	
P10	E minor	41		41		41		41		41
P11	F major	18		18	18		18			18
P12	F minor	22		22		22		22	22	
P13	F# major	30		30		30		30	30	
P14	F# minor	24		24	24			24		24
P15	G major	19		19		19	19			19
P16	G minor	19		19		19	19		19	
P17	A♭ major	44	44			44		44		44
P18	G# minor	29		29	29			29		29
P19	A major	24	24			24		24	24	
P20	A minor	28		28		28		28		28
P21	B♭ major	20		20		20		20		20
P22	B♭ minor	24		24		24		24	24	
P23	B major	19	19			19	19			19
P24	B minor	47	47		47			47		47

(continued)

9.8 Preludes and Fugues Separately

Table 9.14 (continued)

	Index		Pattern 5		Pattern 6		Pattern 7	
			17658S	17658C	17698S	17698C	18654S	18654C
	Length		273	546	273	546	273	546
	Count		8	16	8	16	8	16
P1	C major	35		35		35		35
P2	C minor	38		38		38		38
P3	C# major	104	104		104			104
P4	C# minor	39		39		39	39	
P5	D major	35		35		35	35	
P6	D minor	26		26		26		26
P7	E♭ major	70		70		70	70	
P8	E♭ minor	40		40		40		40
P9	E major	24	24		24			24
P10	E minor	41	41			41		41
P11	F major	18	18		18			18
P12	F minor	22		22	22		22	
P13	F# major	30		30		30	30	
P14	F# minor	24	24			24		24
P15	G major	19	19		19			19
P16	G minor	19	19			19		19
P17	A♭ major	44		44		44		44
P18	G# minor	29		29		29	29	
P19	A major	24	24			24		24
P20	A minor	28		28		28	28	
P21	B♭ major	20		20	20		20	
P22	B♭ minor	24	24			24		24
P23	B major	19		19	19			19
P24	B minor	47		47	47			47

Table 9.15 Example Left = Right pattern over 2 layers in WTC1 fugues

	Index		102S	102C	102S_1S	102S_1C	102C_55S	102C_55C
	Layer		1	1	2	2	2	2
	Length		423	846	141	282	282	564
	Count		8	16	2	6	6	10
F1	C major	27	27			27		
F2	C minor	31	31			31		
F3	C# major	55	55		55			
F4	C# minor	115	115			115		
F5	D major	27		27				27
F6	D minor	44		44			44	
F7	E♭ major	37		37			37	
F8	D# minor	87		87				87
F9	E major	29		29				29
F10	E minor	42		42				42
F11	F major	72		72			72	
F12	F minor	58		58				58
F13	F# major	35	35			35		
F14	F# minor	40	40			40		
F15	G major	86	86		86			
F16	G minor	34	34			34		
F17	A♭ major	35		35				35
F18	G# minor	41		41			41	
F19	A major	54		54			54	
F20	A minor	87		87				87
F21	B♭ major	48		48				48
F22	B♭ minor	75		75				75
F23	B major	34		34			34	
F24	B minor	76		76				76

9.9 Ariadne Musica

Johann Caspar Ferdinand Fischer (1662–1746) also wrote a collection of preludes and fugues in (almost) all keys with the title *"Ariadne Musica"*, probably in 1702 (Fischer). As shown by Tomita (1996), and by Dr. Markus Zepf in a lecture at the Bachfest Leipzig 2018 entitled *"Mit Ariadne's Faden durch das 'Wohltemperirte Clavier'. J. S. Bach studiert Johann Caspar Ferdinand Fischer"* (see also Zepf), Bach may well have taken Fischer's collection as inspiration for WTC1, and this is also mentioned by Jones. It is therefore worth investigating the proportions in this.

Ariadne Musica consists of 20 preludes and fugues in the following keys:

1. C major
2. c# minor
3. d minor
4. D major
5. E♭ major
6. Phrygian
7. e minor
8. E major
9. f minor
10. F major
11. f# minor
12. g minor
13. G major
14. A♭ major
15. a minor
16. A major
17. B♭ major
18. b minor
19. B major
20. c minor

Fischer omits C#/D♭ major, d#/e♭ minor, F#/G♭ major, g#/a♭ minor, a#/b♭ minor and includes one in Phrygian mode.

The solutions and patterns are shown in Table 9.16.

There are 2807 1:1 solutions and this falls in the 99th percentile of random samples (see Fig. 9.17).

In terms of probability, we use the formula from Sect. 8.2.3, $2^{20}/2-1$, for the 20 pieces giving a probability of $2807/524{,}287 = 0.0054$ or 0.54% of a 1:1 proportion occurring by chance. This is somewhat higher than the 0.17% for WTC1.

The situation is similar for the 2:3 proportion. The 4019 solutions fall in the 99.9th percentile of the random samples and there are five symmetrical patterns. The probability of a 2:3 proportion occurring by chance is $4019/2^{20}-1 = 0.0038$ or 0.38%.

We could conclude that Fischer is just as likely as Bach to have used proportions, or that Bach followed Fischer's example. On the other hand, if we believe that it is all due to chance, we would expect to find proportions in Fischer as well.

Table 9.16 Solutions and patterns in Ariadne Musica

Proportion	Solutions	Symmetrical patterns
1:1	2807	1
1:2	–	–
2:3	4019	5

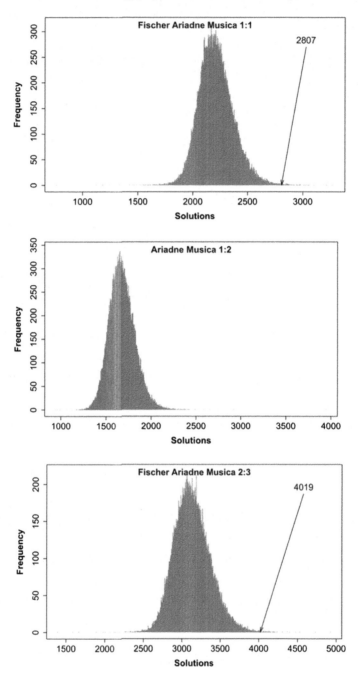

Fig. 9.17 Monte Carlo simulations of Ariadne Musica

9.9 Ariadne Musica

Table 9.17 Summary of Solutions and Patterns in Ariadne Musica

	Ariadne musica one layer			Ariadne musica two layers (Strict)		
	Solutions	Left = Right + Left = thgiR	Proportion count	Solutions	Left = Right + Left = thgiR	Proportion count
1:1	2807	1 + 0	1442	2294	1 + 0	1025
1:2	–	–	–	–	–	–

Second Layer

For a second layer we find 2294 strict solutions, of which 1025 have the 1:1 proportion in the number of pieces (Table 9.17).

There is a parallel 1:1 proportion in the number of major and minor pieces (10:10) if we count Phrygian as minor, as well as a 1:1 proportion in the number of preludes and number of fugues (20:20) given by the nature of the collection.

Chapter 10
Consolidated Observations

10.1 Preamble

Having applied our methods in detail to WTC1 (Chap. 9), I ran the explorer program with several other collections and works. As these do not make for interesting reading, the details have been banished to Appendix A for reference. Some observations made by looking at all of these are given in this chapter. The main statistics are summarised at the end of the chapter in Table 10.22.

We use abbreviated names for the works and collections with the various methods of counting the pieces to obtain better table layouts. The corresponding full names are given here in Table 10.1.

10.2 The Effect of the Number of Pieces

Examining the various collections and works indicated that the number of solutions obtained depends on the number of pieces, and this is no surprise—the more pieces, the more combinations there are and the more solutions for a particular proportion will be found. However, the overall number of combinations also increases exponentially with the number of pieces, so the probability of finding a solution actually decreases as the number of pieces increases.

Tables 10.2 and 10.3 show this for one and two layers of 1:1 proportions and Tables 10.4 and 10.5 show it for one and two layers of 1:2 proportions. These are obtained by running Monte Carlo simulations with 100,000 samples for sets of pieces with lengths in the range from 15 to 168 bars. This range was selected from the range of lengths of all the pieces in all the collections and works considered here, omitting the "outliers", i.e. those shortest and longest pieces that only occur once, to cover 95% of all the lengths Bach used.

The average number of solutions for 1:1 and 1:2 starts to go above 1 with 10 pieces and rises exponentially from there. The values for 24 pieces are different from those

Table 10.1 Short and full names of works and collections

Short name	Full name
Ariadne Musica (Fischer)	Ariadne Musica by Johann Caspar Ferdinand Fischer
Bmin Mass Breve Lutheran	Mass in B minor BWV 232/I, Lutheran version, movements counted at the Breve
Bmin Mass Breve Sections	Mass in B minor BWV 232/II, Catholic version, main sections counted at the Breve
Bmin Mass Breve Movements	Mass in B minor BWV 232/II, Catholic version, movements counted at the Breve
Bmin Mass Semibreve Lutheran	minor Mass BWV 232/I, Lutheran version, movements counted at the semibreve
Bmin Mass Semibreve Sections	Mass in B minor BWV 232/II, Catholic version, main sections counted at the Semibreve
Bmin Mass Semibreve Movements	Mass in B minor BWV 232/II, Catholic version, movements counted at the Semibreve
Brandenburgs Movements	Brandenburg Concertos BWV 1046–1051, movements
Brandenburg Movements all rpts	Brandenburg Concertos BWV 1046–1051, movements with all repeats
Brandenburgs Works	Brandenburg Concertos BWV 1046–1051, the six works as pieces
Canonic Variations	Canonic Variations on "vom Himmel hoch " BWV 769
Cello Suites Movements (AMB)	Cello Suites BWV 1007–1012, Anna Magdalena Bach copy, movements
Cello Suites Works (AMB)	Cello Suites BWV 1007–1012, Anna Magdalena Bach copy, the six works as pieces
CÜ1 Movements	Clavierübung I—Partitas BWV 825–830, movements
CÜ1 Works	Clavierübung I—Partitas BWV 825–830, the six works as pieces
CÜ2	Clavierübung II—Italian Concerto BWV 831 and French Overture BWV 971, movements
CÜ3	Clavierübung III—Organ Mass BWV 669–689, 552, 802–805
CÜ3 Alternative	Clavierübung III—Organ Mass BWV 669–689, 552, 802–805, alternative counting with repeats and time signature change
English Suites Movements	English Suites BWV 806–811, movements
French Suites Movements	French Suites BWV 812–817, movements
French Suites Works	French Suites BWV 812–817, the six works as pieces
Goldbergs	Goldberg Variations BWV 988
Great 15	Great Fifteen Organ Preludes BWV 651–665

(continued)

10.2 The Effect of the Number of Pieces 155

Table 10.1 (continued)

Short name	Full name
Inventions and Sinfonias	Inventions and Sinfonias BWV 772–801
Musical Offering	Musical Offering BWV 1079
Schübler Chorales	Schübler Chorales BWV 645–650, not counting repeats or da capos
Schübler Chorales incl. DC	Schübler Chorales BWV 645–650, including da capos as separate pieces but not repeats
Schübler Chorales plus DC	Schübler Chorales BWV 645–650, counting da capos and repeats in six pieces
Sei Soli Movements	Sei Soli for Violin BWV 1001–1006, movements
Sei Soli Works	Sei Soli for Violin BWV 1001–1006, the six works as pieces
Transcribed Concertos Works with rpts	Transcribed Concertos BWV 972–980, 592, 981–982, movements with repeats
Transcribed Concertos Works wo rpts	Transcribed Concertos BWV 972–980, 592, 981–982, movements without repeats
Trio Sonatas Movements	Trio Sonatas for Organ BWV 525–530, movements without repeats
Trio Sonatas Movements with repeats	Trio Sonatas for Organ BWV 525–530, movements with repeats
Trio Sonatas Works	Trio Sonatas for Organ BWV 525–530, the six works as pieces without repeats
Trio Sonatas Works with repeats	Trio Sonatas for Organ BWV 525–530, the six works as pieces with repeats
Violin Sonatas Movements	Violin Sonatas BWV 1014–1019, movements without repeats
Violin Sonatas Mvmts. repeats	Violin Sonatas BWV 1014–1019,
Violin Sonatas Works	Violin Sonatas BWV 1014–1019, the six works as pieces without repeats
Violin Sonatas Works with rpts	Violin Sonatas BWV 1014–1019, the six works as pieces with repeats
WTC1	The Well Tempered Clavier Book 1 BWV 846–869

given for WTC1 in Sect. 9.2 because we are using the full range of possible lengths as opposed to the range of WTC1. The minimum number of solutions goes above zero from 17 pieces for 1:1, i.e. for more than 16 pieces there is (almost) certainly at least one solution. For 1:2 this happens from 15 pieces.

We see a similar picture for the double proportion with the number of pieces but with lower numbers (PropCounts columns). Only the samples with an even number of pieces can have the 1:1 proportion and those with a multiple of three can have the 1:2 proportion.

Table 10.2 Number of single layer 1:1 solutions against number of pieces

Pieces	1:1 combinations	Min. solutions	Avg. solutions	Max. solutions	Avg. probability bars %	Min. prop counts	Avg. prop counts	Max. prop counts	Avg. probability count %
2	1	0	0.013	1	1.298	0	0.013	1	1.298
3	3	0	0.016	1	0.532				
4	7	0	0.030	3	0.431	0	0.026	3	0.369
5	15	0	0.053	3	0.356				
6	31	0	0.100	4	0.323	0	0.071	4	0.230
7	63	0	0.185	4	0.294				
8	127	0	0.349	6	0.275	0	0.219	4	0.173
9	255	0	0.661	9	0.259				
10	511	0	1.253	10	0.245	0	0.710	8	0.139
11	1023	0	2.380	18	0.233				
12	2047	0	4.573	22	0.223	0	2.373	17	0.116
13	4095	0	8.786	33	0.215				
14	8191	0	16.949	64	0.207	0	8.136	37	0.099
15	16,383	0	32.829	95	0.200				
16	32,767	0	63.568	141	0.194	0	28.603	88	0.087
17	65,535	12	123.386	254	0.188				
18	131,071	21	240.098	455	0.183	0	102.040	218	0.078
19	262,143	186	467.450	949	0.178				
20	524,287	388	911.922	1644	0.174	101	368.000	712	0.070
21	1,048,575	953	1780.595	3038	0.170				

(continued)

10.2 The Effect of the Number of Pieces

Table 10.2 (continued)

Pieces	1:1 combinations	Min. solutions	Avg. solutions	Max. solutions	Avg. probability bars %	Min. prop counts	Avg. prop counts	Max. prop counts	Avg. probability count %
22	2,097,151	2113	3480.133	5890	0.166	555	1340.263	2598	0.064
23	4,194,303	4862	6809.082	11,255	0.162				
24	8,388,607	9544	13,334.709	20,909	0.159	2807	4922.317	8976	0.059

Table 10.3 Number of strict two-layer 1:1 solutions against number of pieces

Pieces	1:1 combinations	Min. solutions	Avg. solutions	Max. solutions	Avg. probability %	Min. propcounts	Avg. propcounts	Max. propcounts	Avg. probability count %
2	1	0	0.000	0	0.000	0	0.000	0	0.000
3	3	0	0.000	0	0.000				0.000
4	7	0	0.000	0	0.000	0	0.000	0	0.000
5	15	0	0.000	3	0.000				0.000
6	31	0	0.000	3	0.000	0	0.000	0	0.000
7	63	0	0.000	3	0.000				0.000
8	127	0	0.000	3	0.000	0	0.000	3	0.000
9	255	0	0.001	5	0.000				0.000
10	511	0	0.003	5	0.001	0	0.000	0	0.000
11	1023	0	0.012	11	0.001				0.000
12	2047	0	0.041	12	0.002	0	0.011	9	0.001
13	4095	0	0.136	23	0.003				0.000
14	8191	0	0.479	34	0.006	0	0.000	0	0.000
15	16,383	0	1.566	47	0.010				0.000
16	32,767	0	5.205	94	0.016	0	1.066	41	0.003
17	65,535	0	16.372	157	0.025				0.000
18	131,071	0	49.695	278	0.038	0	0.000	0	0.000
19	262,143	0	142.858	647	0.054				0.000
20	524,287	0	390.649	1236	0.075	0	89.664	497	0.017
21	1,048,575	0	998.301	2430	0.095				0.000

(continued)

10.2 The Effect of the Number of Pieces

Table 10.3 (continued)

Pieces	1:1 combinations	Min. solutions	Avg. solutions	Max. solutions	Avg. probability %	Min. propcounts	Avg. propcounts	Max. propcounts	Avg. probability count %
22	2,097,151	0	2387.679	5828	0.114	0	0.000	0	0.000
23	4,194,303	0	5389.167	11,500	0.128				0.000
24	8,388,607	0	11,616.536	18,995	0.138	0	3282.622	8991	0.039

Table 10.4 Number of single-layer 1:2 solutions against number of pieces

Pieces	1:2 combinations	Min. solutions	Avg. solutions	Max. solutions	Avg. probability %	Min. propcounts	Avg. propcounts	Max. propcounts	Avg. probability count %
2	3	0	0.018	1	0.594				
3	7	0	0.033	3	0.471	0	0.029	3	0.417
4	15	0	0.055	3	0.367				
5	31	0	0.090	5	0.290				
6	63	0	0.163	5	0.258	0	0.113	5	0.180
7	127	0	0.288	6	0.227				
8	255	0	0.516	7	0.203				
9	511	0	0.929	13	0.182	0	0.519	11	0.102
10	1023	0	1.697	12	0.166				
11	2047	0	3.087	19	0.151				
12	4095	0	5.641	25	0.138	0	2.662	15	0.065
13	8191	0	10.393	47	0.127				
14	16,383	0	19.150	59	0.117				
15	32,767	3	35.388	96	0.108	0	14.537	49	0.044
16	65,535	15	65.421	155	0.100				
17	131,071	43	121.374	342	0.093				
18	262,143	97	225.528	567	0.086	1	82.288	567	0.031
19	524,287	161	419.521	815	0.080				
20	1,048,575	421	781.806	1661	0.075				
21	2,097,151	855	1456.713	2854	0.069	246	477.969	993	0.023

(continued)

10.2 The Effect of the Number of Pieces

Table 10.4 (continued)

Pieces	1:2 combinations	Min. solutions	Avg. solutions	Max. solutions	Avg. probability %	Min. propcounts	Avg. propcounts	Max. propcounts	Avg. probability count %
22	4,194,303	1467	2720.021	5211	0.065				
23	8,388,607	3057	5084.122	9536	0.061				
24	16,777,215	5948	9512.271	17,231	0.057	1740	2830.310	5181	0.017

Table 10.5 Number of two-layer 1:2 solutions against number of pieces

Pieces	1:2 combinations	Min. solutions	Avg. solutions	Max. solutions	Avg. probability %	Min. propcounts	Avg. propcounts	Max. propcounts	Avg. probability count %
2	3	0	0,000	0	0,000				0,000
3	7	0	0,000	0	0,000	0	0,000	0	0,000
4	15	0	0,000	0	0,000				0,000
5	31	0	0,000	0	0,000				0,000
6	63	0	0,000	2	0,000	0	0,000	0	0,000
7	127	0	0,000	2	0,000				0,000
8	255	0	0,001	4	0,000				0,000
9	511	0	0,004	6	0,001	0	0,001	3	0,000
10	1023	0	0,013	8	0,001				0,000
11	2047	0	0,042	15	0,002	0	0,000	0	0,000
12	4095	0	0,128	13	0,003				0,000
13	8191	0	0,392	32	0,005				0,000
14	16,383	0	1,118	31	0,007	0	0,000	0	0,000
15	32,767	0	3,109	62	0,009				0,000
16	65,535	0	8,107	96	0,012				0,000
17	131,071	0	20,610	182	0,016				0,000
18	262,143	0	50,563	295	0,019	0	8,077	98	0,003
19	524,287	0	118,005	568	0,023				0,000
20	1,048,575	0	267,641	911	0,026				0,000
21	2,097,151	0	588,464	1971	0,028	0	0,000	0	0,000

(continued)

10.2 The Effect of the Number of Pieces

Table 10.5 (continued)

Pieces	1:2 combinations	Min. solutions	Avg. solutions	Max. solutions	Avg. probability %	Min. propcounts	Avg. propcounts	Max. propcounts	Avg. probability count %
22	4,194,303	0	1271,726	3692	0,030				0,000
23	8,388,607	213	2701,083	8047	0,032		0,000		0,000
24	16,777,215	973	5663,778	17,725	0,034	0		0	0,000

The exponential rise in the number of possible combinations shows why it is impractical to obtain results for the larger collections and works, the number of possible combinations doubles for each piece added, and so does the time needed to test them. This is also the reason why I only went up to 24 pieces—this took nearly a day for two layers—and is sufficient to show the trend.

The way the number of solutions increases exponentially with the number of pieces is shown graphically for the 1:1 proportion in Fig. 10.1 and for 1:2 in Fig. 10.2. The plots on the left side are for single layer and the right side for two strict layers. The upper row shows the number of solutions for 2–24 pieces and the middle row shows the number of double proportions for 2–24 pieces, i.e. the number of solutions that have the proportion in the count of pieces (PropCount) as well as the lengths in bars. These are shown as stacked bar charts with the minimum, average and maximum

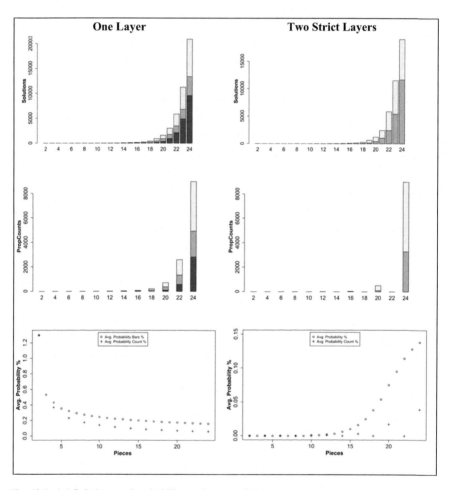

Fig. 10.1 1:1 Solutions and probability against no. of pieces

10.2 The Effect of the Number of Pieces

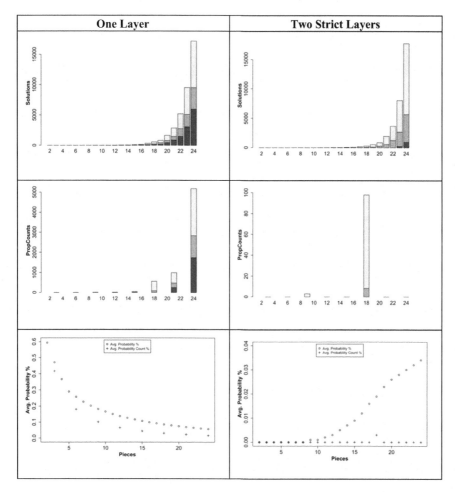

Fig. 10.2 1:2 Solutions and probability against no. of pieces

numbers on top of each other. This was chosen rather than a line graph because the number of pieces is discrete and there are no entries for odd numbers of pieces for 1:1 and only entries for numbers of pieces divisible by three for 1:2.

For a second layer of proportion, the data is very similar. Any sublayer simply consists of a smaller set of pieces taken from the previous layer, so we would expect the number of layer 2 solutions to grow similarly to layer 1, but with a lag. In fact, for 1:1, the minimum number does not go above zero for 24 pieces, and the average goes above one from 15 pieces. For 1:2 these thresholds are 23 and 14 pieces.

This might lead us to expect that the more pieces there are in a set, the easier it is to find a combination. But as we have seen, the number of possible combinations also rises exponentially with the number of pieces, doubling for each piece added. To obtain the probability of finding a solution by chance we need to divide the average

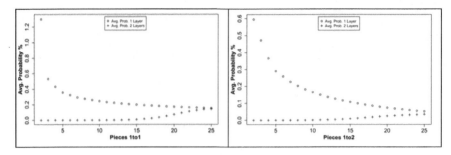

Fig. 10.3 Convergence of probabilities for layers

number of solutions by the number of possible combinations. This is shown in the "Avg. Probability" columns of Tables 10.2, 10.3, 10.4 and 10.5 and graphically in the left plots in the bottom row of Figs. 10.1 and 10.2. For two layers, each layer-1 solution has multiple possibilities at layer 2, so the probability actually increases with the number of pieces (bottom right plots in Figs. 10.1 and 10.2).

In general, we can consider collections at the work level (e.g. 6 sonatas) or at the movement level ($6 \times 3 = 18$ movements). As the above shows, taking collections at the works level with few pieces does not usually give many solutions. Of course, it also depends on the constellation of the lengths as the case of the Goldberg Variations shows (Sect. A.11).

We saw above that the average probability of finding solutions in one layer decreases with the number of pieces, whereas the value for two layers increases. Obviously, if a two-layer solution exists there must also be a one-layer solution, so the probability of two-layer solutions cannot exceed the probability of one-layer solutions. If we plot these values (Fig. 10.3) we see that they converge, so that from about 25 pieces one-layer and two-layer solutions are equally likely. (The above tables were extended to 25 pieces with only 10,000 samples rather than 100,000 to obtain the result in a reasonable time).

> For collections of pieces with random lengths in the range commonly used by Bach, the minimum number of single-layer solutions for the 1:1 proportion increases exponentially with the number of pieces, but the probability of finding a solution by chance decreases.
> With 10 pieces the average number of solutions reaches 1 and so proportions become more likely from there on.
> With more than 17 pieces the minimum number of solutions exceeds 0 and so there will (almost) certainly be proportions.
> For two layers the average number of solutions exceeds 1 from 15 pieces but the minimum not until after 24 pieces, so the presence of proportions is not certain for 24 or fewer pieces. The average probability of finding a strict two-layer proportion increases with the number of pieces. This is because

10.2 The Effect of the Number of Pieces

> each layer 1 solution can have multiple layer 2 solutions given by various combinations of the subsets.

> For the 1:2 proportion the picture is similar. For one layer the average number of solutions reaches 1 with 10 pieces and the minimum at 15 pieces. For two layers the thresholds are at 14 and 23 pieces.

> The decreasing average probability of finding solutions with one layer converges with the increasing probability of finding solutions with two layers. They converge to about 0.15% for 1:1 and 0.044% for 1:2.

10.3 Works Which Could Have More Than One Layer

A collection or work can have solutions for layer L and proportion m:n if the total length is divisible by $(m + n)^L$, e.g. for 1:2 a second layer is possible if the total length is divisible by 3^2 or 9. For the double proportion, this must additionally apply to the number of pieces. Whether it is possible for a set to have 1:1 or 1:2 proportions over one or two layers for the bars or number of pieces is shown in Table 10.6.

10.4 Probability

While considering the six sonatas and partitas for solo violin, the movements of which are between 272 and 524 bars long, (Tatlow 2015) p. 140 writes:

> The probability of six terms between 272 and 524 falling randomly into a perfect double 2:1 proportion is minimal. Bach must have planned it.

And on p. 149:

> Without deliberate design, the bar totals of any six works within a range of 272 and 524 bars in any musical collection are highly unlikely to form an exact double 2:1 proportion.

This was the initial trigger for me to extend the explorer program to include Monte Carlo simulation, as this is the tool we need to test these statements. Taking 100,000

Table 10.6 Possible layers for works and collections

Work or Collection	Pieces	Bars	Proportion: 1:1				1:2			
			Layers: 1		2		1		2	
			Bar	Pc	Bar	Pc	Bar	Pc	Bar	Pc
Ariadne Musica (Fischer)	20	620	Y	Y	Y	Y	N	N	N	N
Bmin Mass Breve Lutheran	12	1040	Y	Y	Y	Y	N	Y	N	N
Bmin Mass Breve Sections	5	2345	N	N	N	N	N	N	N	N
Bmin Mass Breve Movements	27	2345	N	N	N	N	N	Y	N	Y
Bmin Mass Semibreve Lutheran	9	1143	N	N	N	N	Y	Y	Y	Y
Bmin Mass Semibreve Sections	5	2640	Y	N	Y	N	Y	N	N	N
Bmin Mass Semibreve Movements	24	2640	Y	Y	Y	Y	Y	Y	N	N
Brandenburgs Movements	22	2500	Y	Y	Y	N	N	N	N	N
Brandenburg Movements all rpts	25	2808	Y	N	Y	N	Y	N	Y	N
Brandenburgs Works	6	2500	Y	Y	Y	N	N	Y	N	N
Canonic Variations	5	166	Y	N	N	N	N	N	N	N
Cello Suites Movements (AMB)	48	4000	Y	Y	Y	Y	N	Y	N	N
Cello Suites Works (AMB)	6	4000	Y	Y	Y	N	N	Y	N	N
CÜ1 Movements	41	2178	Y	N	N	N	Y	N	Y	N
CÜ1 Works	6	2178	Y	Y	N	N	Y	Y	Y	N
CÜ2	14	942	Y	Y	N	N	Y	N	N	N
CÜ3	27	1840	Y	N	Y	N	N	Y	N	Y
CÜ3 Alternative	27	2055	N	N	N	N	Y	Y	N	Y
English Suites Movements	42	2070	Y	Y	N	N	Y	Y	Y	N
French Suites Movements	35	1380	Y	N	Y	N	Y	N	N	N
French Suites Works	6	1380	Y	Y	Y	N	Y	Y	N	N
Goldbergs	32	960	Y	Y	Y	Y	Y	N	N	N
Great 15	15	1200	Y	N	Y	N	Y	Y	N	N
Inventions and Sinfonias	30	1032	Y	Y	Y	N	Y	Y	N	N
Musical Offering	10	1100	Y	Y	Y	N	N	N	N	N
Schübler Chorales	6	255	N	Y	N	Y	Y	Y	N	N
Schübler Chorales incl. DC	6	280	Y	Y	Y	N	N	Y	N	N
Schübler Chorales plus DC	8	280	Y	Y	Y	Y	N	N	N	N

(continued)

10.4 Probability

Table 10.6 (continued)

Work or Collection	Pieces	Proportion: Bars	1:1 Layers: 1 Bar	Pc	2 Bar	Pc	1:2 Layers: 1 Bar	Pc	2 Bar	Pc
Sei Soli Movements	32	2400	Y	Y	Y	Y	Y	N	N	N
Sei Soli Works	6	2400	Y	Y	Y	N	Y	Y	N	N
Transcribed Concertos Works with rpts	12	3150	Y	Y	N	Y	Y	Y	Y	N
Transcribed Concertos Works wo rpts	12	2764	Y	Y	Y	Y	N	Y	N	N
Trio Sonatas Movements	18	1560	Y	Y	Y	N	Y	Y	N	Y
Trio Sonatas Movements with repeats	18	1860	Y	Y	Y	N	Y	Y	N	Y
Trio Sonatas Works	6	1560	Y	Y	Y	N	Y	Y	N	N
Trio Sonatas Works with repeats	6	1860	Y	Y	Y	N	Y	Y	N	N
Violin Sonatas Movements	25	1790	Y	N	N	N	N	N	N	N
Violin Sonatas Mvmts. repeats	25	2400	Y	N	Y	N	Y	N	N	N
Violin Sonatas Works	6	1790	Y	Y	N	N	N	Y	N	N
Violin Sonatas Works with rpts	6	2400	Y	Y	Y	N	Y	Y	N	N
WTC1	24	2088	Y	Y	Y	Y	Y	Y	Y	N

random sets of six lengths between 272 and 524 which can give a 1:2 proportion (i.e. whose total is divisible by 3) we obtain the result in Table 10.7.

Table 10.7 Monte Carlo simulation of Sei Soli Works 1:2

Solutions	L1_Bars	L1_Pieces
0	93,411	93,411
1	6425	6425
2	125	125
3	39	39

Looking first at the proportions in the numbers of bars, the average number of solutions found in a sample is 0.068[1] (by far the most samples have no solutions). For the six works, there are $2^6-1 = 63$ possible combinations.

> The probability of finding an exact 1:2 proportion in a collection of six works with lengths between 272 and 524 bars is therefore $0.068/63 = 0.0011$ or 0.11%.

This is less than the 0.258% given in Table 10.4 because we are using a smaller range of lengths—that of the specific work rather than the range of all the works.

The simulation ignored 198,960 samples which cannot give a 1:2 proportion (as we would expect, only about one third of the generated pseudo-random samples has a total length divisible by 3). Taking these into account, the probability is 0.00035 or 0.035%, one third of the above. However, there is no point in trying lengths that cannot work, and a composer attempting to achieve a proportion would not consider these, so we will not either.

Turning now to the solutions that also have the proportions in the numbers of pieces—the quotes above from (Tatlow 2015) refer to a "double 2:1 proportion, meaning a 2:1 proportion in the numbers of bars as well as in the numbers of pieces— one would expect the double proportion to be less probable than the single proportion in the numbers of bars alone. But we see that the "L1_Pieces" column in Table 10.7 has the same figures as the "L1_Bars" column. This means that all the samples that had solutions in the numbers of bars also had the proportion in the numbers of pieces. Therefore, the probability of finding the double proportion is the same as the probability of finding the single proportion. I was suspicious of this and thought that it might be due to having many duplicate sets in the 100,000 random samples of six lengths, but in fact there were no duplicates at all—this is easily checked by using the Excel "remove duplicates" function on the results file produced by the program. The length histogram (Fig. 10.4) shows that the pseudo-random lengths are sufficiently uniform.

Re-running the test several times gave the same result with slightly varying numbers as would be expected, but always the same result for the double proportion. I conclude that this is due to the small number of pieces combined with the relatively small range of lengths. We can see from Sect. 10.2 that the two proportions of the double diverge with increasing numbers of pieces, and if we use six pieces while halving the lower limit of the length, i.e. six pieces between 136 and 524 bars, we do see the expected difference—see Table 10.8. More samples have zero solutions for the number of pieces and for those samples that have solutions for the bar proportions, less have the proportion in the numbers of pieces.

Some further analysis of this collection can be found in Sect. A.16.

---

[1] There is a total of $(1*6425) + (2*125) + (3*39) = 6792$ solutions found in 100,000 samples, an average of $6792/100000 = 0.06792$. The explorer program delivers this result directly.

10.4 Probability

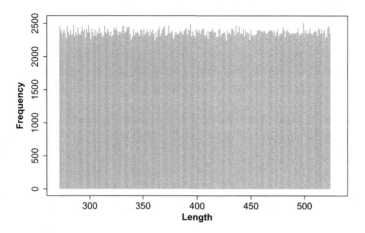

Fig. 10.4 Lengths Histogram of Sei Soli 1:2

Table 10.8 Monte Carlo frequencies for six pieces between 136 and 524 bars for 1:2

Solutions	L1_Bars	L1_Pieces
0	95,201	95,625
1	4708	4300
2	80	64
3	10	11
4	1	0

Regardless of how you verbalise your probabilities these values could indeed be called "minimal" or "highly unlikely", i.e. under 10%—see Fig. 7.1). So Tatlow's statements above are supported by this analysis.

Tatlow (2015) also correctly points out that the probability of two collections, the Sei Soli and the Violin Sonatas, both having a 1:2 proportion is even less—it is $0.0011 \times 0.0018 = 0.0000019$ or 0.00019%. (See Table 10.22 for the figures.) This is further explored in Sect. 10.13.

In another statement for the individual work BWV 1006 with seven movements, (Tatlow 2015) p. 138 states:

As numerous combinations of seven random numbers between 24 and 138 can create a 1:1 proportion, this result with 194:194 bars could easily be dismissed as arithmetical coincidence.

Again, we can test this with the Monte Carlo simulation, and this shows that with 100,000 samples in the above range, there are between 0 and 6 solutions with a mean of 0.21, which for the possible combinations of 7 pieces, $2^7/2-1 = 63$ is a probability of $0.21/63 = 0.00337$ or 0.34%. But this is only for the proportion in the numbers of bars. This again supports Tatlow's statement that they may have been planned, i.e. not coincidence.

Note that as opposed to the 1:2 case above, for the 1:1 proportion the combinations are divided by two to avoid counting the opposites of the solutions as explained in Sect. 8.2.3.

10.5 Types of Distribution

Looking at the overall results of the Monte Carlo simulations, we see three distinct types of distribution. The tables show the sets to which each applies.

- **Similar to Normal or Gaussian** (Table 10.9)
  These are what we would expect. Many of them are not strictly normal, having a slightly longer tail at the right-hand side (some higher numbers of solutions occurring with low frequencies).
- **Too Few Samples** (Table 10.10)
  These are simulations that were run with 1000 samples rather than 100,000, too few samples for the central limit theorem to take effect. They would become more similar to Normal distributions as more samples were simulated (see Sect. 7.5).
- **Few Pieces** (Table 10.11)
  This shape of histogram occurs with a small number of pieces in the work or collection, ten or less, where most samples have no solutions. This corresponds with our expectation from Table 10.2, where we can see that the average number of single-layer 1:1 solutions reaches one between 9 and 10 pieces and the minimum reaches one between 16 and 17 pieces.
  The majority of samples have no solutions and the maximum number of solutions is below 10.

It would be possible to use curve fitting to determine which distribution best fits the data, but I have not pursued that here.

The entries in the following tables apply to the tested proportions 1:1 and 1:2 unless explicitly stated otherwise.

10.5 Types of Distribution

Table 10.9 Collections or works with distributions similar to normal

Collection or work	No. pieces	Example
Ariadne Musica	20	
Mass in B minor Semibreve Movements	24	
Brandenburgs	22	
Brandenburgs rpts	25	
Clavierübung II	41	
Clavierübung III	27	
Clavierübung III Alternative	27	
Great 15 Organ Preludes	15	
Inventions & Sinfonias	30	
Schübler Chorales DC 1:2	8	
Transcribed Concertos Works	12	
Transcribed Concertos Works Repeats	12	
Trio Sonatas Movements	18	
Trio Sonatas Movements Repeats	18	
Violin Sonatas Movements 1:1	25	
Violin Sonatas Movements Repeats 1:1, 1:2	25	
WTC1	24	

Table 10.10 Collections or works with too few samples

Collection or work	Example
French Suites (1000 Samples)	
Goldbergs (1000 Samples)	
Sei Soli Movements (1000 Samples)	
Violin Sonatas Movements (1000 Samples)	

Table 10.11 Collections or works with few pieces

Collection or work	No. pieces	Example
Mass in B minor SB Main Sections	5	
Canonic Variations	5	
Cello Suites AMB Works	6	
CÜ I Works	6	
English Suites Works	6	
French Suites Works	6	
Musical Offering	10	
Schübler Chorales	6	
Schübler Chorales DC 1:1	8	
Sei Soli Works	6	
Transcribed Concertos Works	12	
Trio Sonatas Works	6	
Trio Sonatas Works Repeats	6	
Violin Sonatas Works	6	
Violin Sonatas Works Repeats	6	

> There are three distinct types of distribution for the Monte Carlo simulations. A normal-like distribution is only obtained if there are enough pieces and enough samples are taken.

10.6 Real Works Versus Single-Layer Simulations

When considering the single-layer proportions in the lengths, it is noteworthy that the number of solutions for the actual collections or works is in the 90 percentiles, often the 99th, of the simulated equivalent random sets.

It is also interesting that of the eight simulations with only 1000 samples, none of them achieved the number of solutions of the real work. This is not surprising if we look back at Fig. 7.4 where we see that the highest numbers of solutions, which are also the least likely, only start occurring with a large number of samples. This means that these collections or works are also in the high 90 percentiles of the random sets. The Inventions and Sinfonias 1:2 proportion is an exception here, as the Monte Carlo

10.6 Real Works Versus Single-Layer Simulations

simulation with 100,000 samples still did not include the number of solutions for the real collection.

The histograms are shown in Figs. 10.5, 10.6, 10.7 and 10.8. When looking at these, remember what we said about percentiles in Sect. 9.4: the 99th percentile is 99% of the area under the curve, not 99% of the way along the x-axis.

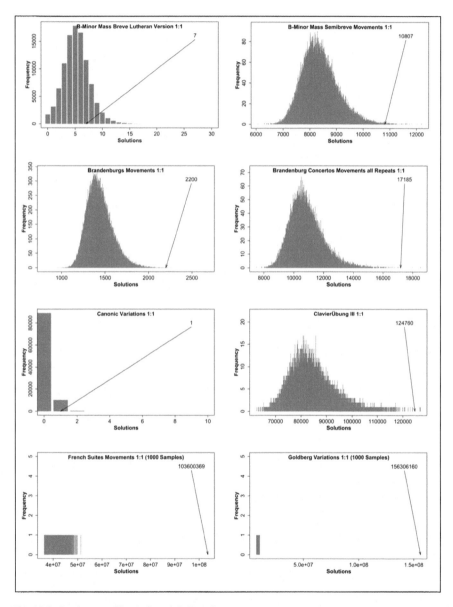

Fig. 10.5 Real versus Simulations 1:1 (Part 1)

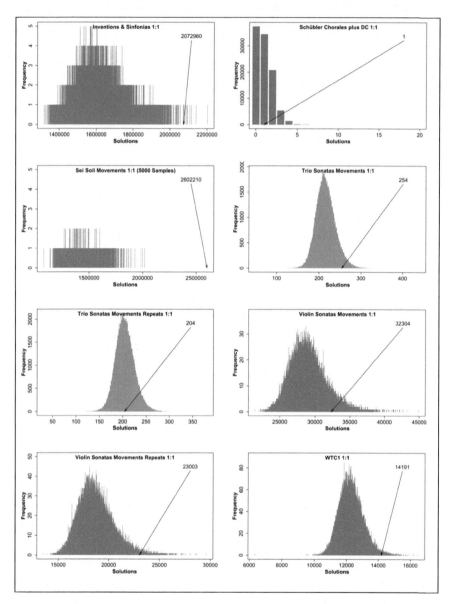

Fig. 10.6 Real versus Simulations 1:1 (Part 2)

The only collections which are below the 90th percentile are:
For the 1:1 proportion (Figs. 10.5 and 10.6):

- Mass in B minor (Lutheran) at the Breve (73.04%)
- Canonic Variations (89.16%)
- The Great 15 Organ Preludes (84.62%)

10.6 Real Works Versus Single-Layer Simulations

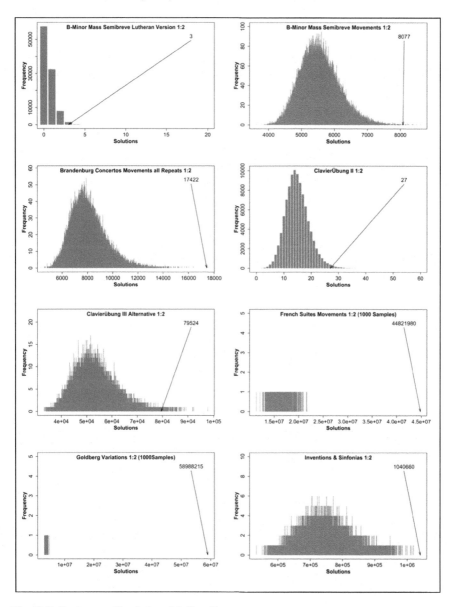

Fig. 10.7 Real versus Simulations 1:2 (Part 1)

- Schübler Chorales with da capos (37.4%)
- Transcribed Concertos with repeats (22.66%) and without repeats (32.32%)
- Trio Sonatas Movements with repeats (50.19%).

For the 1:2 proportion (Figs. 10.7 and 10.8):

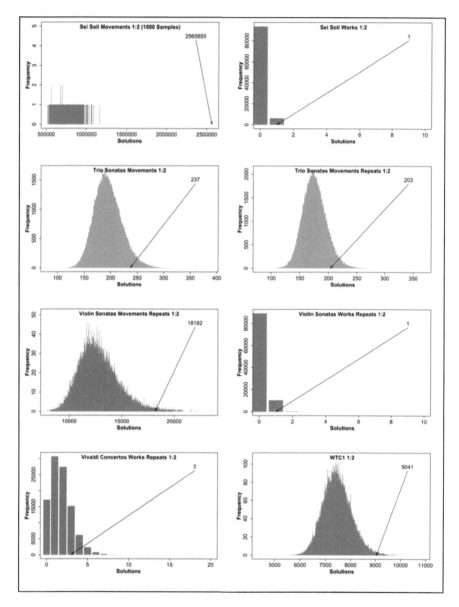

Fig. 10.8 Real versus Simulations 1:2 (Part 2)

- Transcribed Concertos works with repeats (75.16%)
- Trio Sonatas Movements with repeats (89.84%)
- Violin Sonatas works with repeats (89.09%).

Of the pieces that can have the proportion and have solutions, the following are on the 90th percentile or above:

10.6 Real Works Versus Single-Layer Simulations

- For 1:1—12 of 19 (63%)
- For 1:2—13 of 16 (81%).

And if we take the 80th percentile

- For 1:1—14 of 19 (74%)
- For 1:2—15 of 16 (94%).

The complete set of results is shown in Table 10.22 at the end of the chapter.

> An unexpectedly large number of the collections and works have numbers of solutions at the very high end of the Monte Carlo simulations. I can think of no obvious explanation for this, although my story bias (see Sect. 12.2) makes me feel that there should be one. It is tempting to think that the lengths of the pieces in most of Bach's works are chosen to give a relatively large number of solutions for these proportions. This is not plausible, as it is no easier to select the numbers in this way than it is to find the desired proportion directly.

10.7 Accuracy

There are various ways of counting the bars in a piece of music. The main difference is in whether to count repeats and da capo sections or not. This is effectively the difference between the written bars and the sounding bars. If we believe that the proportions can be perceived while listening, at the most subconsciously, then of course the sounding bars with repeats and da capos are the ones to use. I can believe that jazz musicians instinctively follow the 12-bar sections in traditional blues, maybe more from the key changes than by counting, but I am not convinced that anyone can consciously be aware of sections of hundreds of bars. We should ignore the performance practice of playing without the repeats, or only playing the repeat of the first half of a movement but not the second, as this probably originated with trying to fit a complete work on a small number of long-playing vinyl records or compact discs. Another question arises with some suites where a minuet is played again after the subsequent movement, usually either a second Minuet or a Trio, the reprise usually being played without its repeats. This is even more complex in the first Brandenburg Concerto where the Minuet is played and then repeated after Trio I, after the Polacca and after Trio II. If one counts without repeats, an additional problem is the different final bar for the first and second times—they are both written out, but the first-time bar is then not played.

An anacrusis, a partial lead-in bar at the beginning of a piece, can also cause a problem. Strictly, this should be compensated by a partial final bar which taken together give a complete bar and so can be counted as such. However, there are cases where the missing part of the anacrusis is not compensated leaving the question of whether to count it.

Another rare problem arises from time signature changes if they occur without a bar line.

For pieces written in the "old style" stile antico or alla breve, the bars can also be counted at the semibreve, and there is often a small half-bar line in the middle of these bars in the manuscripts.

Tatlow (2015) is exemplary in always pointing out how she is counting.

As far as smaller differences are concerned, we can draw parallels with the study of architectural proportions. Here, researchers are testing whether a building was designed by the architect to have certain proportions, e.g. in width to height, in distances between columns to their widths, etc. Here there are various potential sources of error, such as how to account for the thickness of walls (which are usually depicted by thin lines on the ground plans), whether the thickness of a wall includes the plaster or only the masonry, errors in measurement by the builders, adjustments having to be made due to conditions found at the site or a previously existing building, or changes of mind by the customer. This is obviously more difficult in architecture, where one could decide to ignore discrepancies of a few centimetres as being within the tolerances of the time or the material, but putting a limit on these is subjective and open to biases (see Chap. 12). In the same way, we might contest that a composer intended a certain proportion in the numbers of bars, but that a copying error, lack of space on the page, a last minute change to the libretto or some other reason caused the composition we find today to be inexact—a kind of noise distorting the received signal that we saw in Sect. 2.4.

An interesting French illustration from 1698 in Blondel (1968) showing the musical proportions being applied to the base of a column is shown in Fig. 10.9.

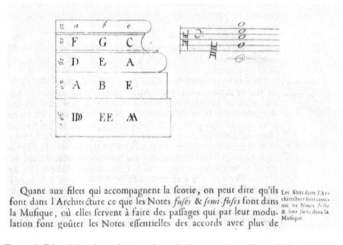

Fig. 10.9 François Blondel, column base and musical proportions. From Cohen, Matthew. Editorial Introduction: Two Kinds of Proportion in (Internet31). Creative Commons CC BY

10.7 Accuracy

A large collection of articles on proportions in architecture can be found in (Internet31) (in which I first saw the above diagram).

The situation is slightly different for our proportions between the lengths in bars of pieces of music. The number of bars in a work or collection is not subject to measurement errors in the same way as measuring distances with a tape measure. Even with a laser that measures to within 1 or 2 mm, the exact positioning of the instrument itself in a cathedral is not repeatable to this precision, and at that size one starts to encounter unevenness in the surface of the stone. A bar of music is not something that can be stretched in this way—either it is counted or it is not counted.

It would be possible to enhance the explorer program to include tolerances and find slightly inexact proportions, a sort of fuzzy search, and this would certainly produce more proportions.

> It is difficult to decide how to count the bars and how to handle the various problem cases. The method chosen should be clearly defined and consistently used. As (Tatlow 2015) says: "The success of historically informed theory is dependent upon accurate data." and "To be able to claim that a composer deliberately incorporated numerical patterns into a work, the data must be pure, based upon the composer's original scores."

10.8 Proportions and Other Structures

We have confirmed that there are various layers of proportion in the collections and works, as (Tatlow 2015) shows. I also expected to find that the proportions relate to other architectural attributes such as in WTC1 the major to the minor or in the Sei Soli the Sonatas to the Partitas.

There are many places in the literature where other structures of works are given. To take one example, Fig. 10.10, which can be found, for example, in Smend (1950), Wolff (1991) (originally 1968) and Leaver (1997) shows a symmetrical structural

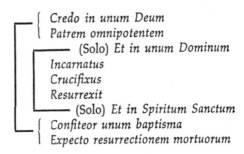

Fig. 10.10 Example structure in Mass in B minor. from Smend (1950) with kind permission of Bärenreiter-Verlag Karl Vötterle GmbH & Co. KG

Table 10.12 Lengths of the pieces in Fig. 10.10

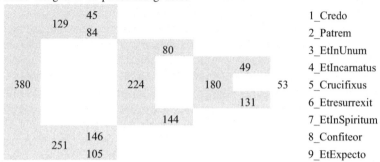

grouping of 1:1 proportions in the Credo of the Mass in B minor around the central point of the Crucifixus.

Looking at the lengths using those given in Tatlow (2015) page 334 to see if there is any correspondence with this structure, we find none—see Table 10.12.

I have not checked any other examples.

Perhaps this expectation is not justified in the context of proportional thinking in Bach's time. This is in contrast to proportions in architecture where the proportions are between the width and height of a portal, the length and width of a building, the spacing of columns, etc., and symmetry is all-important. Of course, other examples would have to be examined before drawing any conclusions.

10.9 Proportions in Durations

Ulrich Siegele postulates that Bach planned his works to have certain durations rather than a certain number of bars (Siegele 2014–18). Tatlow (2015) p. 113–119 examines the historical evidence for this. Note that Siegele does not believe that obtaining specific durations was the goal of composition, but that the proportions in the durations were a starting point to initially structure a work and were therefore flexible.

The Lutheran church service was carefully timed, with chorales before and after the sermon, which was the main part. There were even sand-glass timers on the pulpit and in the organ loft to ensure the appropriate times were kept (Tatlow 2015). These were available in hour, half-an-hour, a quarter-hour and half a quarter-hour (7½ min). I was able to observe this tradition apparently being continued in the present day—while researching in Leipzig I went to a motet service in Bach's own St. Thomas' Church. This started at six p.m. The first performances of music including a Bach motet lasted exactly 25 min, then came the sermon of exactly 10 min and then a second set of hymns and Mendelssohn motets of 25 min, so that the service lasted exactly one hour with the sermon exactly at the midpoint. I did not see any hourglasses or mobile phones being used as timers! I later asked the minister, but she said that they

10.9 Proportions in Durations

do not check the time while preaching but prepare the sermon to a specific duration based on the number of words. The exact timing I experienced was just chance.

One could use the explorer program to look at the proportions of the durations rather than the lengths. This would be logical if one assumes that the proportions were intended to be heard, even if subconsciously. However, Siegele explicitly bases his analyses on the composed lengths, including da capos but not repeats (Siegele 2014–18, vol. 1 p. 28).

Siegele uses a combination of the time signature and tempo levels (*Tempostufe*) which are in proportions 2: 3: 4: 6: 8: 12. The time signature and the tempo level are combined with the predominant division of the beat to give a "rate of motion" (*Bewegungsgrad*).

When trying to analyse the pieces in this way, the following issues arise:

- The premise of having two tempi in a 3:4 proportion is based on single numbers for durations: Praetorius with a pulse of 80 bars in 7½ min for the slower tempo and Mizler with 105 bars in 7½ min for the faster tempo. These work out to 80 bars and 105 bars respectively for 7½ min of music at a beat of 42.66 and 56 beats per minute. The sources given are 135 years apart, so may not be comparable.
- The above values of 80 and 105 are not exactly in 3:4 proportion, so Siegele uses 81 and 108 giving beats of 43.2 and 57.6. Using Mizler as the basis would have given 78¾ and 105 bars for 7½ minutes with whole number 42 and 56 beats per minute in a 3:4 proportion.
- The fractions from the calculations are truncated rather than rounded in Siegele's Book 1 but rounded up in Book 4.
- Caution is needed in obtaining the duration for pieces where the time signature changes.
- The bar numbers used are not the same as (Tatlow 2015)—this type of problem was already mentioned in Sect. 6.2.
- With the first example of the Goldberg Variations, the overall structure assumed is not quite symmetrical. The differences are justified by assuming they are intentional to form a kind of dovetailing or interlocking between the structural units. We will return to this in another context in Sect. 13.2.
- Although it is postulated that Bach was aiming at specific overall lengths, e.g. 45 min or 90 min, these are (not surprisingly) only achieved approximately. We can relate this to the equivalent problem in architectural proportions from Sect. 10.7.

The above should not be taken as criticism of Siegele's method. He explained to me in an email that the use of Praetorius and Mizler was just a starting point, and that the numbers he uses are not intended as arithmetic quantities but a shorthand for the fundamental musical and compositional reality of the durations. This is also the reason why he did not worry about consistency in the roundings.

Due to the fact that the 57.6 beats per minute used as the basis is not an integer, the durations in minutes and seconds are not whole numbers and therefore not suitable for finding proportions with the program. Therefore, the number of normalised bars was used, i.e. the bars converted to 4/4 time taking the tempo level and rate of motion into account, and truncated. The values for these given by Siegele as the basis for his

Table 10.13 WTC1 preludes & fugues durations proportions

	WTC1 Siegele Durations	
	Solutions	Left=Right + Left=thgiR
1:1	–	–
1:2	12,510	4 + 4

Table 10.14 Goldberg Variations durations proportions

	Goldberg Siegele Durations		
	Solutions by Length	Solutions by Duration	Left=Right+ Left=thgiR
1:1	156,306,160	54,288,056	620 + 650
1:2	58,988,215	26,019,978	570 + 587

durations were used (Siegele, 2014–18) vol. 3 p. 73 for the preludes and p. 71 for the preludes using the autograph lengths for consistency).

The values for WTC1 with a total normalised length of 1563 bars only give 1:2 proportions for the 24 preludes and fugues shown in Table 10.13. The preludes alone sum up to a prime number, so have no proportions, and the fugues alone could have 1:2 proportions.

Using the lengths from Tatlow (2015) for a normalised total of 1573 bars instead of Siegele's to calculate the durations cannot give any simple proportions below 1:10 for the 24 pieces.

The Goldberg Variations, which have a very large number of solutions due to the lengths nearly all being the same (see Sect. A.11), have more diverse durations, and therefore less solutions. This is shown in Table 10.14, which includes the solutions by length in the first column for comparison.

In conclusion, the durations seem to give less possibility of finding proportions, but there are too many variables in calculating them to find a meaningful result.

Looking for proportions in durations has similar problems to proportions in architecture that we saw in Sect. 10.7. Not only is it difficult to decide the tolerance one would use (accurate to how many seconds should the timing be?), but each performance will be slightly different.

> The use of durations rather than lengths in bars brings up additional problems in deciding which method to use for obtaining proportions.

10.10 Works with No Proportions

Table 10.15 Collections and works with no solutions

1:1		1:2	
Work	Pieces	Work	Pieces
Bmin Mass Semibreve Sections	5	Bmin Mass Semibreve Sections	5
Brandenburgs Works	6		
Cello Suites Works (AMB)	6		
CÜ1 Works	6	CÜ1 Works	6
CÜ2	14		
French Suites Works	6	French Suites Works	6
Musical Offering	10		
Schübler Chorales incl. DC	6	Schübler Chorales	6
Sei Soli Works	6		
Trio Sonatas Works	6	Trio Sonatas Works	6
Trio Sonatas Works with repeats	6	Trio Sonatas Works with repeats	6
Violin Sonatas Works	6		
Violin Sonatas Works with rpts	6		

Using the explorer program to find proportional 1:1 or 1:2 solutions in the lengths shows up several works or collections that do not have any solutions even though it would be possible to divide the total length into a proportion. The list is easily obtained by filtering the summary Table 10.22 in Excel to obtain Table 10.15 (see Table 10.1 for the full names of the works):

This is neither surprising nor significant. Sets with as few as 6 pieces would not be expected to have proportions, although some of the above do. Cases where there is no proportion at the work level do have proportions at the movements level or do have proportions if the bars are counted differently.

I have performed Monte Carlo simulations on these and as expected, the corresponding random collections do have proportions.

10.11 The Impossible Proportions

The following collections or works have lengths which cannot be divided up into a given proportion. 1:1 is impossible with an odd number of bars and 1:2 is impossible if the total length is not divisible by 3.

Table 10.16 shows the works for which the proportions are impossible, depending on the counting.

Table 10.16 Collections and works with no possible solutions

1:1		1:2	
Work	Pieces	Work	Pieces
Bmin Mass Semibreve Lutheran	9	Bmin Mass Breve Lutheran	12
Bmin Mass Breve Sections	5	Bmin Mass Breve Sections	5
Bmin Mass Breve Movements	27	Bmin Mass Breve Movements	27
		Brandenburgs Movements	22
		Brandenburgs Works	6
		Canonic Variations	5
		Cello Suites Mvmts. (AMB)	48
		Cello Suites Works (AMB)	6
CÜ3 Alternative	27	CÜ3	27
		Musical Offering	10
Schübler Chorales	6	Schübler Chorales incl. DC	6
		Schübler Chorales plus DC	8
		Transcribed Concertos Works wo rpts	12
		Violin Sonatas Movements	25
		Violin Sonatas Works	6

The conclusion is the same as above, although the above proportions are not possible, other proportions are there.

Taking these together, if we look at a collection either as a small number of works or a larger number of movements, and if we consider counting repeats or not:

> All the collections and works tested here have either a 1:1 or a 1:2 proportion in one way or another.

10.12 Reverse Engineering and the Art of Fugue BWV 1080

> **Reverse Engineering**
>
> A well-known technique in computer programming is "reverse engineering", which means finding out the underlying design of a computer program, either by testing it or examining the code. This is used if the design documentation has been lost or was never written, or if the source code has been lost.

10.12 Reverse Engineering and the Art of Fugue BWV 1080

Table 10.17 Versions of the Art of Fugue

		1:1		1:2	
Version	No bars	No solutions	No patterns	No solutions	No patterns
Original Print (Tatlow 2015) P. 244	2,329	–	–	–	–
With chorale	*2,374*	*12,112*	*4*	–	–
With 41 empty bars	2,370	5,839	7	4,640	13
With Tovey	2,402	5,674	1	–	–
With Göncz	2,462	5,270	8	–	–
Later version Wolff/Peters	2,089	–	–	–	–
With chorale	*2,134*	*1,576*	*3*	–	–
With Rechsteiner	2,222	656	0	–	–
With Göncz	2,200	679	0	–	–

Another possible application of the Proportional Parallelism Explorer program is using the possible solutions for proportions to reverse engineer works that are incomplete or exist in several different versions, or to validate reconstructions of lost works. If one assumes that a proportion such as 1:1 or 1:2 was intended, then a version of the work that gives such a proportion would be preferred over one that does not. As we have seen in Sect. 10.2, any collection or work with more than 15 pieces will almost certainly give 1:1 proportions, so this will not help. But we have also seen that the real works tend to lie in the high percentiles of the Monte Carlo simulations, so we might be justified in preferring these.

For example, the Art of Fugue exists in an earlier and later version, and various attempts have been made to complete the unfinished final fugue Contrapunctus 14. There are also opinions that it was intentionally left unfinished, or that the chorale published in its stead should be counted—these are summarised in Table 10.17. Table 10.18 gives details of the individual pieces. [Not all these versions are treated by Tatlow (2015)].

The printed edition contained the 45-bar chorale "Wenn wir in höchsten Nöten". This is normally considered to be irrelevant as it was added by C.P.E. Bach and Altnickol in the published version as a kind of compensation for the incompleteness of the work. Bach dictated this version (BWV 668a) on his deathbed at the same time as he was trying to finish the Art of Fugue. As far as we know, he did not explicitly give instructions that this be added to the Art of Fugue, so it may be far-fetched to consider it, but looking at the rows in italics in Table 10.17, including it provides over twice as many 1:1 solutions as other options.

If we dismiss the idea that it was Bach's intention to include the chorale, the other results shown could be said to support Tatlow's idea that the final fugue was intentionally left unfinished with 41 (= JSBach in the numeric alphabet) bars missing, as this gives the most 1:1 solutions with the second highest number of symmetrical

Table 10.18 Reverse engineering example art of Fugue

Kunst der Fuge (original print—Tatlow p. 244)		Kunst der Fuge original print Tatlow + 41		Kunst der Fuge Tovey (s. Tatlow p. 239, note 55)		Kunst der Fuge original print Tatlow + Göncz		Kunst der Fuge (later version Wolff / Peters)		Kunst der Fuge Rechsteiner p. 61		Kunst der Fuge Göncz cp14 completion	
Cp1	78	Cp1	78	Cp1	78	Cp1	78	Cp1	78	Cp1	78	Cp1	78
Cp2	84	Cp2	84	Cp2	84	Cp2	84	Cp2	84	Cp2	84	Cp2	84
Cp3	72	Cp3	72	Cp3	72	Cp3	72	Cp3	72	Cp3	72	Cp3	72
Cp4	138	Cp4	138	Cp4	138	Cp4	138	Cp4	138	Cp4	138	Cp4	138
Cp5	90	Cp5	90	Cp5	90	Cp5	90	Cp5	90	Cp5	90	Cp5	90
Cp6a4Fr	79	Cp6a4Fr	79	Cp6	79	Cp6	79	Cp6a4Fr	79	Cp6	79	Cp6	79
Cp7a4AugDim	61	Cp7a4AugDim	61	Cp7	61	Cp7	61	Cp7a4AugDim	61	Cp7	61	Cp7	61
Cp8a3	188	Cp8a3	188	Cp8	188	Cp8	188	Cp8a3	188	Cp8	188	Cp8	188
Cp9a4Duod	130	Cp9a4Duod	130	Cp9	130	Cp9	130	Cp9a4Duod	130	CaAug	109	CaAug	109
Cp10a4Dec	120	Cp10a4Dec	120	Cp10	120	Cp10	120	Cp10a4Dec	120	Cp9	130	Cp9	130
Cp11a4	184	Cp11a4	184	Cp11	184	Cp11	184	Cp11a4	184	CaDec	82	CaDec	82
Cp12/1Inva4	56	Cp12/1Inva4	56	Cp12	56	Cp12	56	Cp12/1Inva4	56	CaDuo	78	CaDuo	78
CP12/2Inva4	56	CP12/2Inva4	56	Inva4	56	Inva4	56	CP12/2Inva4	56	Cp10	120	Cp10	120
Cp13/1Inva3	71	Cp13/1Inva3	71	Cpa3	71	Cpa3	71	Cp13/1Inva3	71	CaOtt	103	CaOtt	103
Cp13/2Inva3	71	Cp13/2Inva3	71	Inva3	71	Inva3	71	Cp13/2Inva3	71	Cp11	184	Cp11	184
Cpa4	98	Cpa4	98	Cpa4	98	Cpa4	98	CaOtt	103	Cp12_1	56	Cp12_1	56
CaAug	109	CaAug	109	CaAug	109	CaAug	109	CaDecTerz	82	CP12_2	56	CP12_2	56
CaOtt	103	CaOtt	103	CaOtt	103	CaOtt	103	CaDuoQuint	78	CP13_1	71	CP13_1	71
CaDec	82	CaDec	82	CaDec	82	CaDec	82	CaAugContr	109	CP13_2	71	CP13_2	71
CaDuo	78	CaDuo	78	CaDuo	78	CaDuo	78	Cp14	239	CP14Rek	372	CP14Rek	350
Fua2	71	Fua2	71	Fua2	71	Fua2	71						

(continued)

10.12 Reverse Engineering and the Art of Fugue BWV 1080

Table 10.18 (continued)

Kunst der Fuge (original print—Tatlow p. 244)		Kunst der Fuge original print Tatlow +41		Kunst der Fuge Tovey (s. Tatlow p. 239, note 55)		Kunst der Fuge original print Tatlow + Göncz		Kunst der Fuge (later version Wolff / Peters)	Kunst der Fuge Rechtsteiner p. 61	Kunst der Fuge Göncz cp14 completion
AlMo	71	AlMo	71	AlMo	71	AlMo	71			
Fu3Sog	239	Fu3Sog	280	Fu3Sog	312	Cp14Göncz	372			2200
	2329		2370		2402		2462	2089	2222	
Choral	45							45		
	2374							2134		

patterns and is the only one with 1:2 solutions, also with patterns. Of course, this would also support completions of 41 bars.

The number of patterns refers to "Left=Right" and "Left=thgiR" (mirror image) only.

There is also a completion by Davitt Moroney bringing Contrapunctus 14 to 269 bars, but the total length does not have any proportions with this.

Kevin Korsyn is also working on completions,[2] but informed me that there are various versions with different numbers of bars, so they are not included here.

This could be applied to other works for which there is no definitive version, e.g. the Cello Suites (Sect. A.5), Clavierübung III (Sect. A.8), etc.

> If we believe that Bach used proportions, then reconstructions which give a proportion can be favoured over those that do not.

10.13 Combining Works

Taking this quote from Butt (1997):

> In some ways, it may well be productive to consider his entire oeuvre as a single musical work, of which the individual instances are—literally—*pieces* ('Stücke').

We might try and find proportions over related collections or even the entire oeuvre. Tatlow (2015) considers this as well, for example showing an overarching scheme for the 24 Preludes and Fugues of WTC1 interlocking with the 15 Inventions and Sinfonias (p. 168–169) and going on to include the Clavierübung series (p. 202). But as she points out, for many works we do not have a definitive final edition from which to take the lengths, many works are lost, and we cannot be certain which works should be included.

Looking within the works, it is obvious that if the individual works or collections have an internal proportional structure, then works or collections taken together will have the same proportion: if the ratios a:b and c:d and e:f are equal, then $(a + c + e):(b + d + f)$ is the same ratio.

Another way of considering the total output is to consider the overall probability that all the main collections or works have 1:1 or other proportions by chance. As the individual probabilities are small, this becomes smaller the more collections and works we take into consideration. For example, the probability that the four in Table 10.19 all have a 1:1 proportion by chance is about $5 \times 10.^{12}$, obtained by multiplying the individual probabilities.

---

[2]For example, forthcoming: Imagining Fragment X: Completing Bach's Art of Fugue (Oxford University Press).

10.13 Combining Works

Table 10.19 Combining probabilities for multiple works

1:1	No. pieces	No. solutions	No. combinations	Probability	Monte Carlo average No. solutions	Probability of Monte Carlo average
WTC1	24	14,191	8,388,607	0.17%	12,400	0.15%
6 Soli Movements	32	2,602,210	2,147,483,647	0.12%	1,428,975	0.07%
6 Sonatas Movements	25	32,304	16,777,215	0.19%	18,282	0.11%
Bmin Mass Movements Semibreve	24	10,807	8,388,607	0.13%	28,826	0.34%
Probability of all				$5 \times 10.^{12}$		$4 \times 10.^{12}$

This is well below the threshold of significance used by particle physicists (see Sect. 7.3).

An additional visualisation of the data is shown in Fig. 10.11. For all the 535 pieces or movements in all the 19 collections and works considered here, the number of pieces (frequency) having each of the lengths is shown. There is a noticeable peak at a length of 32 bars, mainly because 26 of the Goldberg Variations have this length, and another at 24 bars mainly from the English and French Suites. The third highest frequency is 48 bars, mainly from the Cello Suites and English and French Suites. The range beyond about 200 bars is sparsely populated, and the longest pieces are the Prelude of the fifth Cello Suite BWV 1011 including repeats at 446 bars, followed by

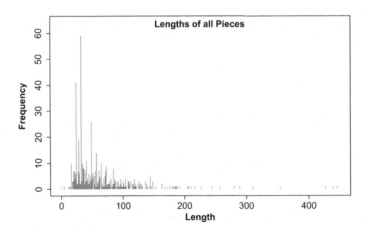

Fig. 10.11 Lengths of Bach's pieces and movements

the Sonata from the Musical Offering BWV 1079 at 440 bars and the first movement of Brandenburg Concerto No. 4 with 427 bars.

This is a rather messy distribution with several peaks and outliers, not in the least bit Gaussian, and is probably neither useful nor interesting.

- Art of Fugue
- Mass in B minor
- Brandenburg Concertos
- Canonic Variations
- Cello Suites
- Clavierübung I
- Clavierübung II
- Clavierübung III
- Clavierübung IV
- English Suites
- French Suites
- Inventions and Sinfonias
- Musical Offering
- Schübler Choral Preludes
- Sonatas for Violin and Harpsichord
- Trio Sonatas for Organ
- Violin Sonatas and Partitas
- Well Tempered Clavier Bk. 1
- Well Tempered Clavier Bk. 2.

WTC1 with Inventions and Sinfonias

Looking in more detail at the combined collections of WTC1 with the Inventions and Sinfonias, the 54 pieces take too long to run full analyses to completion, but for example, exploring the 1:1 proportion with two layers gives no strict solutions when allowed to run for 24 h, but did find one in about 5 h with the pieces sorted by length.

As with WTC1, based on their compositional history, (Tatlow 2015) shows a three-layer 1:1 solution in the 39 pieces with a perfect reflective symmetry in layer 1 (see Fig. 10.12). However, this solution is not quite strict, and the program shows that there are no strict three-layer solutions. This solution as found by the program as the 26,904th is shown in Fig. 10.13, transposed for easier comparison with the original. The extra row "RT column" was added manually and shows the column number in Fig. 10.12. The two empty columns in the middle with the index numbers in brackets are the missing layer 3 solution that would have made it strict through all three layers. The row labelled "PropCount" shows the pair that (Tatlow 2015) points out as having the 1:1 proportion in 9:9 pieces (in the "Count" row).

Tatlow (2015) (page 169–170) also shows a 1:2 solution with consecutive interleaved groups from each collection; this is the 117,963,826th of over 120 million 1:2 solutions,[3] of which over 38 million have a parallel 1:2 proportion in the number of

---

[3]Tatlow uses 2:1 rather than 1:2, so this is about the 3 millionth solution of 120 million.

10.13 Combining Works

BWV	Key	Title	Bars	1:1		1:1		1:1		1:1		1:1	
772	C major	Inventio 1	22	22								22	22
773	C minor	Inventio 2	27	27								27		27
774	D major	Inventio 3	59	59									59	
775	D minor	Inventio 4	52	52								52		52
776	E♭ major	Inventio 5	32	32									32	
777	E major	Inventio 6	62	62									62	
778	E minor	Inventio 7	23			23	23				23			
779	F major	Inventio 8	34			34	34			34				
780	F minor	Inventio 9	34			34		34	34					
781	G major	Inventio 10	32			32	32			32				
782	G minor	Inventio 11	23			23		23		23				
783	A major	Inventio 12	21			21	21				21			
784	A minor	Inventio 13	25			25		25		25				
785	B♭ major	Inventio 14	20			20	20				20			
786	B minor	Inventio 15	22			22		22	22					
			[488]											
787	C major	Sinfonia 1	21			21	21				21			
788	C minor	Sinfonia 2	32			32		32	32					
789	D major	Sinfonia 3	25			25	25			25				
790	D minor	Sinfonia 4	23			23		23		23				
791	E♭ major	Sinfonia 5	38			38	38				38			
792	E major	Sinfonia 6	41			41		41	41					
793	E minor	Sinfonia 7	44			44	44				44			
794	F major	Sinfonia 8	23			23		23		23				
795	F minor	Sinfonia 9	35			35		35		35				
796	G major	Sinfonia 10	33	33								33		
797	G minor	Sinfonia 11	72	72								72		
798	A major	Sinfonia 12	31	31								31		31
799	A minor	Sinfonia 13	64	64								64		64
800	B♭ major	Sinfonia 14	24	24								24		24
801	B minor	Sinfonia 15	38	38								38		38
Totals			[544]	516 : 516		258 : 258		129 : 129		129 : 129		258 : 258		129 : 129

Fig. 10.12 1:1 Proportions in Inventions and Sinfonias from Tatlow (2015). Reproduced with permission of the Licensor through PLSclear

pieces, 835 have a left/right translation or reflection symmetry and 249 have both a parallel proportion in the number of pieces and a symmetry. But none of these have a neat division among the pieces as Tatlow's solution. It is not possible to examine all of this amount of output for other patterns—the 11 GB file cannot even be opened in Excel—so one could only search for suspected patterns.

> This shows that the historically informed method of following the composer's thought processes based on the traditions of the time together with some intuition cannot always be bettered by a computerised brute force approach. All

Index	26904S	26904C	26904S_5S	26904S_5C	26904S_5S_1S	26904S_5S_1C	(26904S_5C_1S)	(26904S_5C_1C)	26904C_253S	26904C_253C	26904C_253S_1S	26904C_253S_1C	26904C_253C_1S	26904C_253C_1C
Layer	1	1	2	2	3	3			2	2	3	3	3	3
Sum	516	516	258	258	129	129			258	258	129	129	129	129
Count	12	18	7	5	4	3			9	9	5	4	4	5
PropCount									Y	Y				
RT Column	1	2	9	10	11	12			3	4	8	7	5	6
Inv1	22		22		22									
Inv2	27		27			27								
Inv3	59			59										
Inv4	52		52			52								
Inv5	32			32										
Inv6	62			62										
Inv7		23							23		23			
Inv8		34							34		34			
Inv9		34								34			34	
Inv10		32							32		32			
Inv11		23								23				23
Inv12		21							21		21			
Inv13		25								25				25
Inv14		20							20		20			
Inv15		22								22			22	
Sin1		21							21		21			
Sin2		32								32			32	
Sin3		25							25		25			
Sin4		23								23				23
Sin5		38							38		38			
Sin6		41								41			41	
Sin7		44							44		44			
Sin8		23								23				23
Sin9		35								35				35
Sin10	33			33										
Sin11	72			72										
Sin12	31		31		31									
Sin13	64		64			64								
Sin14	24		24		24									
Sin15	38		38			38								

Fig. 10.13 Tatlow's Multiple 1:1 Proportions of Inventions and Sinfonias as Output by Program

the program does show in this case is that there are no better solutions than Tatlow's.

Sei Soli and Violin Sonatas

Tatlow (2015) also combines the two collections of violin works noting that both have a total of 2400 bars giving a double 1:1 proportion, although the Sei Soli are counted without repeats and the Sonatas for Violin and Harpsichord are counted with repeats. As I confirm in Sects. A.16 and A.19, both collections can be divided into double 1:2 proportions.

Taking the collections together the explorer program finds 5 ways of dividing the 12 works into 1:2 proportions, all having the double proportion in the number of pieces. These are shown in Table 10.20—solution 1C contains all the Violin Sonatas and solution 5C contains all the Sei Soli.

The Monte Carlo simulation shows that most solutions for the proportion in the numbers of bars also have the double proportion in the numbers of pieces. The real solutions are in the 97th percentile in both cases—see Table 10.21.

10.13 Combining Works

Table 10.20 Sei Soli and Violin Sonatas combined 1:2

Index	1S	1C	2S	2C	3S	3C	4S	4C	5S	5C
Layer	1	1	1	1	1	1	1	1	1	1
Sum	1600	3200	1600	3200	1600	3200	1600	3200	1600	3200
Count	4	8	4	8	4	8	r4	8	4	8
PropCount	Y	Y	Y	Y	Y	Y	Y	Y	Y	Y
BWV1001	272			272		272		272		272
BWV1002	408		408			408		408		408
BWV1003	396			396	396			396		396
BWV1004		412		412		412	412			412
BWV1005	524			524		524		524		524
BWV1006		388		388	388		388			388
BWV1014		328		328		328		328	328	
BWV1015		419	419		419			419	419	
BWV1016		397	397		397		397			397
BWV1017		477		477		477		477	477	
BWV1018		403		403		403	403			403
BWV1019		376	376			376		376	376	

Table 10.21 Monte Carlo simulation of Sei Soli and Violin Sonatas combined 1:2

Solutions	Bars	Pieces
0	19,149	19,495
1	32,562	32,684
2	26,800	26,665
3	13,248	13,038
4	5262	5192
5	1958	1923
6	664	649
7	219	217
8	88	87
9	30	30
10	14	15
11	1	1
12	3	2
13	0	0
14	1	1
15	1	1

10.14 Summary of Main Statistics

Table 10.22 summarises the statistics shown above and in Appendix A.

The following are facts derived from the statistics:

(1) The proportions that are possible depend on the factors of the total number of bars.
(2) Any collection or work with a sufficient number of pieces and random lengths in a sufficient range will give some solutions for the simple proportions.
(3) Some works have no solutions for certain proportions that would be possible for their total lengths. These are nearly all small sets for which this is to be expected and they all have a proportion in a different way.
(4) The results depend on how the bars are counted with regard to repeats, da capo sections, anacruses, time signature changes and counting at the semibreve.
(5) Where proportions exist, the probability of finding a solution by chance is very low. (Of course, you can still find solutions by manual methods, as (Tatlow 2015) does).
(6) The shape of the distribution of solutions for random samples depends on the number of pieces and the number of samples. If either is too small the distribution will not be similar to a normal distribution.
(7) The majority of real collections or works have numbers of solutions in the high (80–90) percentiles of equivalent random sets.
(8) Finding proportions in durations is even more problematical than finding them in written lengths measured in bars.
(9) The potentially large amount of data produced by computer can be overwhelming. In this case it is best used to test hypotheses rather than for speculative exploration.
(10) The data neither proves nor disproves that Bach intentionally used proportions.

The main statistics are summarised in Table 10.22.

All this only goes a small part of the way to testing Tatlow's theory of proportional parallelism. We can test the statistics around the occurrence of proportions, layers of proportions in the numbers of bars and layers of proportions in the numbers of pieces, as well as the occurrence of certain patterns or symmetries in the order of the pieces within the proportions. However, it does not say anything about the history of compositional construction and the changes made to works during the compositional process as evidenced by the original manuscripts, or the theological and aesthetic aspects given by the attitudes of the time.

One must also be aware of biases, nicely put by Wittkower (1949) writing about architecture: "It is true, that in trying to prove that a system of proportion has been deliberately applied by a painter, a sculptor or an architect, one is easily misled into finding those ratios which one sets out to find. ... If we want to avoid the pitfall of useless speculation we must look for unmistakable guidance by the artists themselves." Providing this "unmistakable guidance" is one of the achievements of (Tatlow 2015).

10.14 Summary of Main Statistics

Table 10.22 Summary of solution statistics

1:1	No. pieces	No. solutions	No. combinations	Probability	Monte Carlo average No. Solutions	Probability of Monte Carlo average	Percentile of actual no. solutions
Ariadne Musica (Fischer)	20	2,807	524,287	0.54%	2,210	0.42%	99.90%
Bmin Mass Breve Lutheran	12	7	2,047	0.34%	5,242	0.26%	73.04%
Bmin Mass Breve Sections	5	–	15	–	0.011	0.07%	NA
Bmin Mass Breve Movements	27	–	67,108,863	–	101,202	0.15%	NA
Bmin Mass Semibreve Lutheran	9	–	255	–	0.393	0.15%	NA
Bmin Mass Semibreve Sections	5	0	15	0.00%	0.010	0.07%	No solutions
Bmin Mass Semibreve Movements	24	10,807	8388,607	0.13%	8,357	0.10%	99.90%
Brandenburgs Movements	22	2,200	2097,151	0.10%	1,435	0.07%	99.97%
Brandenburg Movements all rpts	25	17,185	16,777,215	0.10%	10,777	0.06%	99.9980%
Brandenburgs Works	6	0	31	0.00%	0.022	0.07%	No solutions
Canonic Variations	5	1	15	6.67%	0.114	0.76%	89.16%
Cello Suites Movements (AMB)	48	Incomplete	140,737,488,355,327	0.47%	Not feasible	/	/
Cello Suites Works (AMB)	6	0	31	0.00%	0.022	0.07%	No solutions
CÜ1 Movements	41	2,235,988,612	1,099,511,627,775	0.20%	Not feasible	/	/

(continued)

Table 10.22 (continued)

1:1	No. pieces	No. solutions	No. combinations	Probability	Monte Carlo average No. Solutions	Probability of Monte Carlo average	Percentile of actual no. solutions
CÜ1 Works	6	0	31	0.00%	0.063	0.20%	No solutions
CÜ2	14	0	8,191	0.00%	13,373	0.16%	No solutions
CÜ3	27	124,760	67,108,863	0.19%	83,265	0.12%	99.9997%
CÜ3 Alternative	27	—	67,108,863	—	83,228	0.12%	NA
English Suites Movements	42	Incomplete	2,199,023,255,551	1.47%	Not feasible	/	/
French Suites Movements	35	103,600,369	17,179,869,183	0.60%	41,836,308	0.24%	**
French Suites Works	6	0	31	0.00%	0.08859	0.29%	No solutions
Goldbergs	32	156,306,160	2,147,483,647	7.28%	9001,066	0.42%	**
Great 15	15	33	16,383	0.20%	27	0.16%	84.62%
Inventions and Sinfonias	30	2072,960	536,870,911	0.39%	1602,112	0.30%	99.99%
Musical Offering	10	0	511	0.00%	0.497	0.10%	No solutions
Schübler Chorales	6	—	31	—	0.497	1.60%	NA
Schübler Chorales incl. DC	6	0	31	0.00%	0.320	1.03%	No solutions
Schübler Chorales plus DC	8	1	127	0.79%	1,003	0.79%	37.40%
Sei Soli Movements	32	2602,210	2,147,483,647	0.12%	1428,975	0.07%	**
Sei Soli Works	6	0	31	0.00%	0.044	0.14%	No solutions
Transcribed Concertos Works with rpts	12	1	2,047	0.05%	1.522	0.07%	22.66%
Transcribed Concertos Works wo rpts	12	2	2,047	0.10%	2.311	0.11%	32.21%

(continued)

10.14 Summary of Main Statistics

Table 10.22 (continued)

1:1	No. pieces	No. solutions	No. combinations	Probability	Monte Carlo average No. Solutions	Probability of Monte Carlo average	Percentile of actual no. solutions
Trio Sonatas Movements	18	254	131,071	0.19%	216	0.16%	93.37%
Trio Sonatas Movements with repeats	18	204	131,071	0.16%	205	0.16%	50.19%
Trio Sonatas Works	6	0	31	0.00%	0.064	0.21%	No solutions
Trio Sonatas Works with repeats	6	0	31	0.00%	0.056	0.18%	No solutions
Violin Sonatas Movements	25	32,304	16,777,215	0.19%	28,826	0.17%	93.09%
Violin Sonatas Mvmts. repeats	25	23,003	16,777,215	0.14%	18,616	0.11%	99.07%
Violin Sonatas Works	6	0	31	0.00%	0.082	0.26%	No solutions
Violin Sonatas Works with rpts	6	0	31	0.00%	0.073	0.23%	No solutions
WTC1	24	14,191	8388,607	0.17%	12,400	0.15%	99.20%
1:2	No. pieces	No. solutions	No. combinations	Probability	Monte Carlo average no. solutions	Probability of Monte Carlo average	Percentile of actual no. solutions
Ariadne Musica (Fischer)	20	–	1048,575		–	0.15%	NA
Bmin Mass Breve Lutheran	12	–	4,095	–	5.966	0.15%	NA
Bmin Mass Breve Sections	5	–	31	–	0.01785	0.06%	NA

(continued)

Table 10.22 (continued)

1:2	No. pieces	No. solutions	No. combinations	Probability	Monte Carlo average no. solutions	Probability of Monte Carlo average	Percentile of actual no. solutions
Bmin Mass Breve Movements	27	–	134,217,727	–	53.347	0.04%	NA
Bmin Mass Semibreve Lutheran	9	3	511	0.59%	0.546	0.11%	98.13%
Bmin Mass Semibreve Sections	5	0	31	0.00%	0.017	0.05%	No solutions
Bmin Mass Semibreve Movements	24	8,077	16,777,215	0.05%	5,535	0.03%	99.97%
Brandenburgs Movements	22	–	4194,303	–	1,210	0.03%	NA
Brandenburg Movements all rpts	25	17,422	33,554,431	0.05%	8,002	0.02%	99.999%
Brandenburgs Works	6	–	63	–	0.035	0.06%	NA
Canonic Variations	5	–	31	–	0.213	0.69%	NA
Cello Suites Mvmts. (AMB)	48	–	281,474,976,710,655	–	Not feasible	/	/
Cello Suites Works (AMB)	6	–	63	–	0.034	0.05%	NA
CÜ1 Movements	41	854,821,792	2,199,023,255,551	0.04%	Not feasible	/	/
CÜ1 Works	6	0	63	0.00%	0.099	0.16%	No solutions
CÜ2	14	27	16,383	0.16%	14,920	0.09%	99.10%

(continued)

10.14 Summary of Main Statistics

Table 10.22 (continued)

1:2	No. pieces	No. solutions	No. combinations	Probability	Monte Carlo average no. solutions	Probability of Monte Carlo average	Percentile of actual no. solutions
CÜ3	27	–	134,217,727	–	52,541	0.04%	NA
CÜ3 Alternative	27	79,524	134,217,727	0.06%	52,513	0.04%	99.87%
English Suites Movements	42	Incomplete	4,398,046,511,103	1.37%	Not feasible	/	/
French Suites Movements	35	44,821,980	34,359,738,367	0.13%	16,216,224	0.05%	**
French Suites Works	6	0	63	0.00%	0.13808	0.22%	No solutions
Goldbergs	32	58,988,215	4,294,967,295	1.37%	3619,221	0.08%	**
Great 15	15	39	32,767	0.12%	28	0.09%	93.73%
Inventions and Sinfonias	30	1040,660	1,073,741,823	0.10%	737,770	0.07%	**
Musical Offering	10	–	1,023	–	0.697	0.07%	NA
Schübler Chorales	6	0	63	0.00%	0.782	1.24%	No solutions
Schübler Chorales incl. DC	6	–	63	–	0.508	0.81%	NA
Schübler Chorales plus DC	8	–	255	–	1.422	0.56%	NA
Sei Soli Movements	32	2565,655	4,294,967,295	0.06%	732,423	0.02%	**

(continued)

Table 10.22 (continued)

1:2	No. pieces	No. solutions	No. combinations	Probability	Monte Carlo average no. solutions	Probability of Monte Carlo average	Percentile of actual no. solutions
Sei Soli Works	6	1	63	1.59%	0.067	0.11%	93.46%
Transcribed Concertos Works with rpts	12	3	2,047	0.15%	1,749	0.09%	75.16%
Transcribed Concertos Works wo rpts	12	–	2,047	–	2,629	0.13%	NA
Trio Sonatas Movements	18	237	262,143	0.09%	195	0.07%	93.33%
Trio Sonatas Movements with repeats	18	203	262,143	0.08%	176	0.07%	89.84%
Trio Sonatas Works	6	0	63	0.00%	0.099	0.16%	No solutions
Trio Sonatas Works with repeats	6	0	63	0.00%	0.088	0.14%	No solutions
Violin Sonatas Movements	25	–	33,554,431	–	18,768	0.06%	NA
Violin Sonatas Mvmts. repeats	25	18,182	33,554,431	0.05%	12,668	0.04%	99.67%
Violin Sonatas Works	6	–	63	–	0.127	0.20%	NA
Violin Sonatas Works with rpts	6	1	63	1.59%	0.115	0.18%	89.09%
WTC1	24	9,041	16,777,215	0.05%	7,449	0.04%	99.71%

10.14 Summary of Main Statistics

Notes on Table 10.22:

See Table 10.1 for the full names of the works and collections in the first column.

A "-" in the No. Solutions column indicates that no solutions are possible for the proportion. There is then "NA" in the last column.

"Incomplete" in the No. Solutions column indicates that it would take too long to run the solution search to completion, so the probability value is an estimate from a limited number of tries.

"Not feasible" in the MC Average No. Solutions column indicates that the Monte Carlo simulation takes too long and is not feasible. There is then a "/" in the other columns.

"No Solutions" in the Percentile of Actual No. Solutions column indicates that there were no actual solutions for that proportion (0 in the No. Solutions column) and so no percentile can be given.

"**" in the Percentile of Actual No. Solutions column indicates that there are too many pieces to perform a Monte Carlo simulation with 100,000 samples. Using 1000 samples, none of them had as many solutions as the actual collection or work. We could assign a value of 100% as the percentile.

Fischer's Ariadne Musica is given purely for completeness.

10.15 Notes for Researchers

> Statistical techniques must be used with the appropriate understanding of their capabilities and interpretation.

> Proportions can nearly always be found. They must be evaluated in their historical context.

> Always make clear how the bars are being counted and give reasons.

Chapter 11
Magic Squares

11.1 The Dieben Rectangle

As mentioned in Sect. 3.2, derivatives of magic squares have been applied to Bach's works since Henk Dieben first published one for the lengths of the preludes and fugues of the Well Tempered Clavier Book 1. Kramer (2000) gives a good history of the Dutch numerologists and goes on to detail and extend the numbers to be found in WTC1 and other works.

A magic square is a matrix of numbers where all the rows and all the columns and the main diagonals have the same sum. These have been known since the earliest of times. There are many variants of these, e.g. as rectangular matrices rather than squares, and less perfect ones where the diagonals have different sums to the rows and columns or where the rows and the columns have different sums, etc. Classical magic squares consist of a sequence of numbers starting with 0 or 1, e.g. 1–9 for a 3 × 3 square. The mathematics of these is well understood; see, for example, (Ollerenshaw 1998).

The lengths in bars of 24 or 48 pieces such as the preludes and fugues of WTC1 are not made up of consecutive numbers, and also cannot form a square, but only rectangles (which may contain squares). The scheme found by Dieben is shown in Fig. 11.1 reproduced from Kramer (2000). This uses the lengths of the preludes and fugues individually in a 4 × 12 rectangle. Every row adds up to 174 and every column to 522. Each of the three 4 × 4 squares adds up to 696. Every top left and bottom right 2 × 2 square in each of the three main squares adds up to 116 and every top right and bottom left 2 × 2 square adds up to 232. The three central 2 × 2 squares (bold) all add up to 522—the same as the columns. The diagonals of the three 4 × 4 squares add up to 1044 (Mäser2000). (Kramer also adds the significance of 29 as the gematria of SDG Soli Deo Gloria.)

To find all the magic squares for a set of numbers with a personal computer by brute force still does not seem feasible. For 48 pieces it would have to try all 1.2×10^{61} permutations. About 4×10^{17} seconds have elapsed since the universe was

[A] Dieben's Rechteck										D3
1.			35	44	40	55		174		
		116					232			
2.	A		19	18	22	115		174		
									696	Rechteck
3.			41	72	24	37		174		
		232					116			[A]
4.			44	75	24	31		174		
5.			19	29	87	39		174		
		116					232			
6.	B		27	41	48	58		174		
									696	696 = 24×29
7.			34	70	40	30		174		
		232					116			
8.			86	42	26	20		174		
9.			35	38	54	47		174		116 = 4×29
		116					232			
10.	C		19	24	104	27		174		174 = 6×29
									696	
11.			87	35	24	28		174		232 = 8×29
		232					116			
12.			76	34	29	35		174		522 = 18×29
			522	522	522	522		T 2088 = 72 × 29		

Fig. 11.1 Dieben's magic square for WTC1 (From Kramer (2000) with kind permission from the author)

created, so even checking 3×10^{43} permutations per second would take a year. I have therefore not attempted to write a program for this.

I did consider using the Proportional Parallelism Explorer program to break the problem down. For example, searching for the 1:11 proportion with a pattern match for any 4 pieces will give the potential rows that add up to 174 (the total of 2088 bars divided by 12 to give the target of 174 for the 1:11 proportion). There are 1636 of these, found in about 4.5 s. One could then use the binary signatures of these to find all the sets of twelve that comprise all the pieces once. There are $^{1636}C_{12} = 7.37 \times 10^{29}$ of these, again intractable, although one could use some short cuts similar to the explorer program. Going the other way, and using the pattern matching to find the four columns, this time with 12 pieces, with a 1:3 proportion, the program found 383,340,167 patterns that have 12 pieces adding up to 522 in about 3.75 days. This is even more intractable, with $^{383340167}C_4 = 9.0 \times 10^{32}$ combinations to test.

Rows or columns can be permuted as this has no effect on the sums if we disregard the diagonals. Kramer (2000) points out that when one magic rectangle has been found, numerous others can be derived by transposing numbers in one row or column and performing a corresponding transposition of numbers with the same difference in another row or column. To get an idea of the extent of these possibilities, I created a matrix of the lengths of the 48 preludes and fugues to obtain all the differences in the lengths—a part is shown in Fig. 11.2. The leading diagonal (each piece and itself) and the half below this (the pairs of pieces in reverse order) are omitted.

There are several pieces with the same length—more than initially meets the eye—see Table 11.1.

11.1 The Dieben Rectangle

	A	B	C	D	E	F	G	H	I	J	K	L	M	N	O	P	Q	R	S	T	U	V
			P1	F1	P2	F2	P3	F3	P4	F4	P5	F5	P6	F6	P7	F7	P8	F8	P9	F9	P10	F10
			35	27	38	31	104	55	39	115	35	27	26	44	70	37	40	87	24	29	41	42
P1	35			8	-3	4	-69	-20	-4	-80	0	8	9	-9	-35	-2	-5	-52	11	6	-6	-7
F1	27				-11	-4	-77	-28	-12	-88	-8	0	1	-17	-43	-10	-13	-60	3	-2	-14	-15
P2	38					7	-66	-17	-1	-77	3	11	12	-6	-32	1	-2	-49	14	9	-3	-4
F2	31						-73	-24	-8	-84	-4	4	5	-13	-39	-6	-9	-56	7	2	-10	-11
P3	104							49	65	-11	69	77	78	60	34	67	64	17	80	75	63	62
F3	55								16	-60	20	28	29	11	-15	18	15	-32	31	26	14	13
P4	39									-76	4	12	13	-5	-31	2	-1	-48	15	10	-2	-3
F4	115										80	88	89	71	45	78	75	28	91	86	74	73
P5	35											8	9	-9	-35	-2	-5	-52	11	6	-6	-7
F5	27												1	-17	-43	-10	-13	-60	3	-2	-14	-15
P6	26													-18	-44	-11	-14	-61	2	-3	-15	-16
F6	44														-26	7	4	-43	20	15	3	2
P7	70															33	30	-17	46	41	29	28
F7	37																-3	-50	13	8	-4	-5
P8	40																	-47	16	11	-1	-2
F8	87																		63	58	46	45
P9	24																			-5	-17	-18
F9	29																				-12	-13
P10	41																					-1
F10	42																					
P11	18																					

Fig. 11.2 Part of WTC1 differences matrix

Table 11.1 Pieces in WTC1 with the same length

Pieces	No. of pieces	Length
P01, P05, F13, F15	4	35 bars
P09, P14, P19, P22	4	24 bars
P08, F14	2	40 bars
F08, F20	2	87 bars
F09, P18	2	29 bars
P10, F18	2	41 bars
P15, P16, P23	3	19 bars
F16, F23	2	34 bars

Matching up the positive and negative differences, of the 91 distinct differences ranging from −88 to +97, 73 of them (80%) have a corresponding difference of the opposite sign. There is an average of 6.8 corresponding pieces of opposite difference and a maximum of 30. However, to compensate for a simple swap of two pieces in one row, the two compensating pieces must be in the same columns in another row (conversely for columns), making it considerably more difficult. Of course, one can also work with a sequence of swaps. This is somewhat similar to the game of sudoku (see side panel), but more difficult as one has to consider the sums rather than the mere presence of the numbers. A difference of zero is treated separately, as these pairs can be swapped without changing the sums, and only has an effect on the title ordering; there are 22 of these. The distribution is shown in Fig. 11.3. This all makes it seem possible that other magic rectangles could be derived.

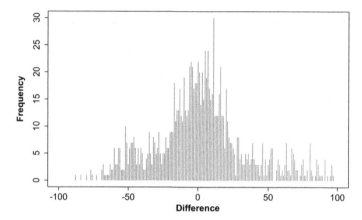

Fig. 11.3 Distribution of WTC1 length differences

Fig. 11.4 Two-layer 1:1 proportion on columns derived from Dieben

35	44	40	55
19	18	22	115
41	72	24	37
44	75	24	31
19	29	87	39
27	41	48	58
34	70	40	30
86	42	26	20
35	38	54	47
19	24	104	27
87	35	24	28
76	34	29	35
522	522	522	522
1044		1044	

11.1 The Dieben Rectangle

> Sudoku has a grid of 9x9 squares. Each row, each column and each of the nine 3x3 rectangles must contain all the numbers 1-9.

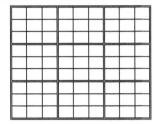

But Bach would probably have only thought of the "genuine" magic squares with consecutive numbers starting from 1 and would not necessarily have thought of arranging his works in arbitrary rectangular forms. He could have known of magic squares. They are treated in Athanasius Kircher's Arithmologia of 1665 with examples of magic squares for the planets: Saturn 3×3, Jupiter 4×4, Mars 5×5, Sol 6×6, Venus 7×7, Mercury 8×8, Luna 9×9. Incidentally, this book also treats numerical alphabets, as (Tatlow 1991) shows.

Although we cannot currently give a probability of Dieben's scheme occurring by chance, as we can see from the above, the likelihood does increase if the pieces already have some proportional properties.

There is apparently no system or order behind the scheme of pieces in Dieben's rectangle. If one is to believe that Bach adjusted the lengths of the pieces to fit into such a scheme, one might think that there would be some sense in the order, and at least the prelude and fugue pairs would be kept together. On the other hand, if the parallel proportions in Chap. 9 are intentional, these do not have any such scheme either. No magic rectangle of the lengths of the prelude and fugue pairs has been shown.

If Bach had created magic squares, they probably would have been squares rather than rectangles. The only collection whose number of pieces is a square is the Violin Sonatas with 25 pieces.

11.1.1 Deriving Proportions from Dieben's Rectangle

Having failed to use the explorer program to find magic rectangles, we can go the other way and use Dieben's rectangle to find multiple layers of proportion. This actually enables us to find proportions in the 48 pieces of the preludes and fugues of WTC1 taken separately, which we did not attempt in Sect. 9.8 as it would have taken too long. For example:

Fig. 11.5 Two-layer 1:1 proportion on rows derived from Dieben

35	44	40	55	174		
19	18	22	115	174	522	
41	72	24	37	174		1044
44	75	24	31	174		
19	29	87	39	174	522	
27	41	48	58	174		
34	70	40	30	174		
86	42	26	20	174	522	
35	38	54	47	174		1044
19	24	104	27	174		
87	35	24	28	174	522	
76	34	29	35	174		

- Take any pair of the four columns to give the first layer 1:1 proportion (1044:1044) and each column of the pair gives the second layer (522:522). See Fig. 11.4. Both layers have a further 1:1 layer in the number of pieces.
- Take any six of the twelve rows to give the first layer 1:1 proportion (1044:1044) and any pair of three rows from them to give the second layer (522:522). Both layers have a further 1:1 layer in the number of pieces. See Fig. 11.5.
- See Fig. 11.6. Take the diagonals (blue lines over grey cells) of the three 4 × 4 squares (black outline) for the solution and the rest (yellow) for the complement to give a 1:1 proportion (1044:1044). For the second layer of the solution use the central 2 × 2 squares (red outline) and the four corner pieces (orange outline) of each 4 × 4 square. For the second layer of the complement use the horizontal pairs of pieces (green outline) and the vertical pairs of pieces (purple outline). Both layers have a further 1:1 layer in the number of pieces.
- See Fig. 11.7. Take one (green) of the three 4 × 4 squares and the other two (orange) for the first layer of a 1:2 proportion (696:1392). Use the diagonally opposing 2 × 2 squares within each (blue, magenta) to give the second layer 1:2 proportion (232:464 and 464:928). Both layers have a further 1:2 layer in the number of pieces.

11.1 The Dieben Rectangle

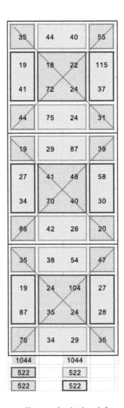

Fig. 11.6 Two-layer 1:1 proportion on diagonals derived from Dieben

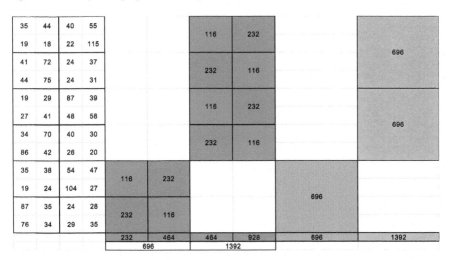

Fig. 11.7 Two layers of 1:2 proportion derived from Deiben rectangle

These can all be derived in various ways by taking different pairs of rows or columns and by transposing the rows or columns and/or transposing pieces as described above.

11.2 Use by Modern Composers

Magic squares have been used in modern times to generate musical ideas. Cross (2003) points out that the composer Peter Maxwell Davis used a magic square to derive *Ave maris stella* and *A mirror of whitening light* in the 1970s. This is reminiscent of the use of dice and concentric wheels to generate ideas for motifs in the eighteenth century, as we shall see in Sect. 13.4. Taruskin (2010) mentions the use of magic squares by the composer Pierre Boulez.

Chapter 12
Psychological Fallacies

12.1 Preamble

While exploring the music and the numbers and looking for proportions, possible messages, statistical significances, etc., one should be aware of certain psychological biases or fallacies that cause human beings to interpret meaning or intention where none exists. These are due to the way in which the human brain attempts to make sense of or derive information from the raw data it perceives from the world.

A lot of the research in this field was done by Kahnemann and Tversky in the field of economics and how real people make investment decisions which are incompatible with the purely rational thinking which many economic models assume.

I do not claim to be immune to these myself, and indeed, it has been shown that awareness of these fallacies does not prevent one from being subject to them.

There have been numerous cases of data being selectively massaged or suppressed to give the desired results, and there is also a tendency to favour the simplest or most pleasing mathematical models to explain the universe. This often occurs subconsciously and unintentionally. This quote from a physicist in Hossenfelder (2018) is particularly apt: "There are things that cry out for an explanation, like a 2-to-1 ratio… When you see that, you feel like you have to explain that." More on this in Chap. 13.

Most of these are described in Dobelli (2011) and a good overview with links to further detail is in (Internet32).

12.2 Story Bias or Narrative Fallacy

One of these is the narrative fallacy described in Kahnemann (2011) and Taleb (2007). This refers to the tendency to force some kind of logical explanation onto the perceived data, even though it may just be random. To quote from Kahnemann (2011):

Narrative fallacies arise inevitably from our continuous attempt to make sense of the world. The explanatory stories that people find compelling are simple; are concrete rather than abstract; assign a larger role to talent, stupidity, and intentions than to luck; and focus on a few striking events that happened rather than on the countless events that failed to happen.

Or Taleb (2007):

> The narrative fallacy addresses our limited ability to look at sequences of facts without weaving an explanation into them, or, equivalently, forcing a logical link, an arrow of relationship upon them. Explanations bind facts together. They make them all the more easily remembered; they help them make more sense. Where this propensity can go wrong is when it increases our impression of understanding.

Or (Internet33):

> However, events will almost never be perceived intuitively as being random; one can find an apparent pattern in almost any set of data or create a coherent narrative from any set of events.

This is related to the phenomenon of apophenia, the tendency to see connections between unrelated things or to discover patterns in random data or noise.

It is easy to see that coming across one or more interesting occurrences of numbers, one can embed these into the narrative of Bach the great genius and extend it from his musical genius into mathematical genius (see also Sects. 12.5 and 12.6).

12.3 Confirmation Bias

This bias is a tendency to selectively use or pay attention to information that confirms one's preconceptions, assumptions or desired outcomes, and to ignore information that contradicts them.

For those hoping to find hidden messages, any numerical coding found in Bach's music will be seen as confirming that Bach actually used such codings or that there is some intentional meaning there, whereas any information to the contrary, e.g. statistics or the absence of such codings where one might expect them, are ignored or not mentioned. The equivalent is also true of those hoping not to find something. Hopefully the information given in the previous chapters may help to counteract this.

12.4 Neglect of Probability

Other research, e.g. by Kahnemann and Tversky, has shown that humans have a poor intuitive understanding of probability and randomness. Even Leibniz knew this—in a letter to Nicolas Remond on 10 January 1714 he writes:

> Elle serviroit aussi à estimer les degrés de vraisemblance, lorsque nous n'avons pas sufficientia data pour parvenir à des vérités certaines, et pour voir ce qu'il faut pour y suppléer.

> Et cette estime seroit des plus importantes pour l'usage de la vie, et pour les délibérations de pratique, où en estimant les probabilités on se mécompte le plus souvent de plus de la moitié.
>
> When we lack sufficient data to arrive at certainty in our truths, it would also serve to estimate degrees of probability and to see what is needed to provide this certainty. Such an estimate would be most important for the problems of life and for practical considerations, where our errors in estimating probabilities often amount to more than a half.
>
> (Leibniz 1714, p. 68)

One commonly quoted example (here from Fenton 2013) is the likelihood that at a party of 23 people, any two would have the same birthday (day and month). Most people assume this is quite unlikely and are amazed when it happens—after all there are 365 possible birthdays in a year. However, it turns out that the probability is 51%, and that you only need to have 20 people for the probability of any two sharing a birthday to be 50%, and in a class of 40 the probability is 90%.

In the same way, if someone discovers that some attribute of a composition by J. S. Bach can be interpreted as 14, and this is also a coding of the word "Bach", then they will tend to assume that this is intentional and has some significance.

On the other hand, even if an event has a very low probability of happening, as long as it is above zero it can still happen. You might consider 1% (or 0.01) to be a very low probability, but on average, the event will happen once in every 100 trials or occur in one of every 100 samples in the long run. For undesirable outcomes, according to Murphy's Law, which is popular among engineers, if anything can go wrong, it will go wrong, but belief in the truth of this is due to confirmation bias (Sect. 12.3).

12.5 Halo Effect

This effect causes us to transfer a person's traits or abilities from one area to another, e.g. by assuming that as Bach was a genius as a composer and musician he is also a genius at mathematics and coding.

I would add a variant of this together with confirmation bias, which I will call "Hero Bias", whereby people tend to glorify the person in question and try to show that he or she is superior in all ways, and so invent achievements or capabilities which are not founded on fact. This, together with confirmation bias, also causes them to ignore or suppress any negative traits or deeds: for example, the fact that Bach spent time in prison is rarely mentioned in early biographies.

As Tatlow (2015, p. 368) says: "Are these claims mere empty rhetoric fuelled by a hagiographic desire to raise Bach to the level of an evangelist?"

It can also work the other way, where for someone perceived as evil, every action will be seen as evil. An example of this is Richard Wagner (born in Leipzig as it happens), whose antisemitism has caused controversy about whether his music should be played in Israel.

12.6 Conjunction Fallacy

This fallacy causes us to think that something is more probable when extra but irrelevant information is given. The widely quoted classical "Linda" example from Kahneman and Tversky is:

> Linda is 31 years old, single, outspoken, and very bright. She majored in philosophy. As a student, she was deeply concerned with issues of discrimination and social justice, and also participated in anti-nuclear demonstrations.
>
> Which is more probable?
> 1. Linda is a bank teller.
> 2. Linda is a bank teller and is active in the feminist movement.

Most people select answer 2, but 1 is the more probable—there are more bank tellers in total than bank tellers who are also feminists. Answer 2 narrows down the possibilities from the overall set, making it less probable.

Applying this here might read:

Bach was the greatest composer of counterpoint and fugal music. He was a member of the Mizler Society of Musical Science. Which is more probable?

1. Bach composed music.
2. Bach composed music and hid numerical messages in it.

12.7 Notes for Researchers

> Various biases and fallacies cause us to assume meaning where there is merely chance and to selectively view evidence that supports our prejudices.

> Being aware of these biases does not necessarily prevent us from being subject to them.

Chapter 13
Bach, Science and Technology

13.1 Mathematics and Philosophy in Bach's Time

In the first half of the eighteenth century, during Bach's life until he died in 1750, there were many treatises on mathematics and much progress in statistics such as the normal distribution, standard deviation, and crossovers between mathematics, philosophy and music, which at that time were not regarded as separate subjects. Seeing the large amount of activity and progress made in probability and statistics up to 1750, we may ask to what extent Bach may have been aware of the chances involved in forming paragrams or proportions. A selection of these events is shown in relation to Bach's life in a tabular timeline in Appendix C.

The history of paragrams and permutation poetry is thoroughly treated in Tatlow (1991). One main thread of influence can be traced through (Lull 1274), Christian Wolff and Mizler (Göncz 2013).

Leibniz (1646–1716) was born in Leipzig and studied at the university there (see Fig. 13.1).

> Leibniz (as mentioned in Tatlow (1991)) also dreamed of inventing a symbolic language for scientific discourse with which one could handle philosophical problems with mathematical accuracy, where errors in logic or false conclusions could be found and corrected, and thus, instead of having discussions to resolve a disagreement, one could say: "Let us calculate". (This well-known quote appears in several of Leibniz' writings, e.g. (Leibniz 1687, 1688).) Hence the motto of this book given on the flyleaf does not simply mean "let us use a computer" but also "let us think logically". Like Gödel in 1931 (see Sect. 2.8), Leibniz also intended to use prime numbers to denote logical concepts and perform operations on them (Antognazza 2016).

Fig. 13.1 Statues of Gottfried Wilhelm Leibniz and Johann Sebastian Bach in Leipzig. Photographs © by the author

This was taken up again in 1912 when Russell and Whitehead followed Leibniz in formulating a complete logical derivation of all mathematical truths in their *Principia Mathematica*, and in 1931 Gödel proved that this can never be complete.

The philosopher Christian Wolff (1679–1754) applied mathematical thinking to philosophy, was a follower of Leibniz and incorporated Leibniz's work (although not the combinatorics) in his *Anfangsgründe aller mathematischen Wissenschaften* (Foundations of all Mathematical Sciences). Wolff was in Leipzig between 1703 and 1706, in Halle (about 35 km from Leipzig) between 1706 and 1723 and in Halle again from 1740 until his death.

Lorenz Christoph Mizler was a self-avowed Pythagorean "apostle of the Wolffian philosophy" (Felbick 2012), and a pupil of Bach between 1731 and 1734. He published various works following Wolff's lead in applying mathematical logic to music and was the founder of the "corresponding society of musical sciences" which Bach joined in 1747. Mizler was in Leipzig from 1731 until 1743 when he moved to Poland, but still visited after that. Mizler developed his theories following Wolff's philosophy with the methods of mathematical proofs, and titled some of his publications similarly to Wolff, i.e. *Anfangsgründe aller musikalischen Wissenschaften nach mathematischer Lehrart abgehandelt* (Foundations of all Musical Sciences Treated According to Mathematical Teaching) and *Anfangsgründe des Generalbaß*

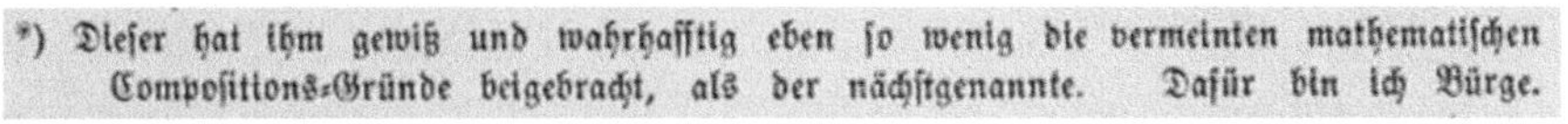

13.1 Mathematics and Philosophy in Bach's Time

*) Dieser hat ihm gewiß und wahrhafftig eben so wenig die vermeinten mathematischen Compositions-Gründe beigebracht, als der nächstgenannte. Dafür bin ich Bürge.

Fig. 13.2 Mattheson's footnote on Bach and Mizler. From Mattheson (1740) https://archive.org/details/grundlageeinereh00matt

nach mathematischer Lehrart abgehandelt (Foundations of Thoroughbass Treated According to Mathematical Teaching. Mizler (1739).

Euler extended Pythagoras to a theory of music and harmony in 1730.

Johann Christoph Gottsched (1700–1766) was at Leipzig University from 1724 and wrote the texts to some of Bach's cantatas. He was a friend of Matthias Gesner (1691–1761), the rector of Mizler's school in Ansbach, was in Weimar at the same time as Bach, and then rector of the St. Thomas school in Leipzig from 1730 to 1734.[1] Both were disciples of Wolff, and Gesner was the president of the "Deutsche Gesellschaft" (a literary society) founded by Gottsched in 1727. One of the statutes of Mizler's society was that articles should be written according to the style of the Deutsche Gesellschaft and Wolffian principles (Tatlow 2015, p. 100–101).

As noted in Göncz (2013), combinations and permutations occur from the earliest cabbalistic texts, and then in (Lull 1274) in the 13th–14th century, permutations of words were used in literature and poetry to generate texts as early as the year 330 AD and were popular in the 17th century, Leibniz uses the invention of musical melodies by performing permutations on the notes of the scale as an example of the application of permutations, and (Mizler 1739) takes this up as well, referring directly to Leibniz's *Ars Combinatoria* in his *Anfangsgründe des Generalbasses* with reference to using permutations for the purpose of musical invention. (Mattheson 1739) talks about permutations, saying how many there are for 1, 2, 3 and 4 things. Bach also composed permutation fugues, cyclically alternating the entries of the themes in the four voices.

Mattheson (1748) published a systematic theory of sound.

This was a time when knowledge was not specialised and thinkers applied themselves to everything, mathematics, geometry, astronomy, philosophy, law, music, literature and poetry.

So, Bach was embedded in a circle of people with wide-ranging mathematical, scientific, literary, theological and musical interests closely associated with the university. Even though Bach himself was not an academic, he could have had access to any of this knowledge. This view is supported by Wolff (2001) (p. 310). If he had asked any of these people: "How many ways are there of dividing up 24 preludes and fugues in a one to one proportion?", they would have been able to give him the answer from Leibniz: $2^{24}-1 = 16{,}777{,}215$.

On the other hand, there is no evidence that Bach had any interest in these matters (see also Sect. 8.1).

There is the well-known footnote by Mattheson in Mizler's entry in the Ehren-Pforte (Mattheson 1740) (Fig. 13.2) saying that Bach certainly did not teach Mizler

---

[1] Bach was in Leipzig from 1723 until his death in 1750.

the presumed mathematical basics of composition any more than he (Mattheson) did.

There is a letter (Suchalla 1994) from Carl Philipp Emanuel Bach to Forkel dated 13 January 1775, where he appears to answer questions posed by Forkel while writing Bach's biography, but unfortunately Forkel's letter with the questions is lost. This says that Bach had no enthusiasm for "dry mathematical stuff" ("Der seelige war, wie ich und alle eigentlichen Musici kein Liebhaber, von trocknem mathematischen Zeuge"). Another statement by Forkel (BD) saying that even the well-versed in mathematics Johan [sic] Sebastian Bach oriented himself in these questions with nature and not the rules, and all the mathematicising has not even been successful in guaranteeing a perfect temperament ("Selbst der in der Mathematik so gelehrte Johan Sebastian Bach habe sich in diesen Fragen nach der Natur, nicht nach der Regel gerichtet, und die ganze Mathematisiererei habe noch nicht einmal den Erfolg gehabt, die Durchführung einer einwandfreien Temperatur zu gewährleisten.") would appear to indicate that Bach did not apply mathematics to tuning but oriented himself naturally, i.e. by ear.

We might therefore conclude that Bach did know some mathematics but did not apply it to composition or tuning. There is no record of Bach having possessed any books of a mathematical or scientific nature (Bayreuther 2013).

Butt (1997) sees Bach's approach to music and composition as similar to Leibniz' and Wolff's, but not directly influenced by them.

A possible area for future research might be to investigate what contacts Bach had with the university in Leipzig, and to what extent he might have been acquainted with the above concepts.

13.2 Models

A model is a representation of some aspects of a thing in another medium. This may be physical such as a scaled down wooden model of a cathedral to enable those commissioning it to visualise its final appearance and select among competing architects, or for the architects themselves to visualise how it can be constructed. Or a model may be an abstract representation as in a drawing.

Models are used extensively in the design of buildings and the engineering of technical systems, especially computer software which is by nature very abstract. These range from simple flow charts to structured design and object-oriented models. These aim to visualise how a complex system is organised and how the various interactions between the parts occur. These models are themselves complex and usually created with the aid of computer programs which can also check the internal consistency and so avoid many errors in the final program code. The designs for solutions to some commonly occurring problems are also available as design templates called patterns.

Music composition also has designs and patterns, for example a concerto consisting of three movement with fast, slow, fast tempi. Composers usually plan the architecture or structure of a work before composing, with themes and developments etc. and these can be seen in surviving sketches. And as (Tatlow 2015) has

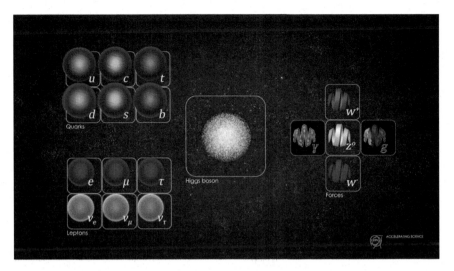

Fig. 13.3 Standard model particles. Image by Dominguez, Daniel, © CERN, used with permission

shown, there is historical and numerical evidence to suggest that Bach drew up plans in which parallel layers of 1:1 and 1:2 proportions characterise the structures of his revised and published collections.

Another use of models is to explain or understand something that already exists, and that is our situation with the music of J. S. Bach written three hundred years ago. Creating models to understand something that already exists is the basis for most sciences such as physics.

Scientists call a possible explanation of the mechanism by which an observed phenomenon occurs a model. When something is observed that cannot be explained by the model, the model is enhanced or replaced by a better one.

In particle physics, the models used to explain the composition and properties of matter started simply and have become ever more complex. First the ancient Greeks thought that if you cut something in half repeatedly you must reach a point beyond which no further reduction is possible and called this the atom. At the beginning of the eighteenth century, Leibniz presented monads as the underlying elements of substance, although not in the physical sense. In 1897, Sir Joseph Thomson discovered the electron. In 1913, Ernest Rutherford and Niels Bohr[2] postulated that the atom was composed of a nucleus with orbiting electrons. In 1919 Rutherford discovered the proton and in 1932 James Chadwick discovered the neutron, components of the nucleus. Since then a whole zoo of even smaller particles has been discovered, the most recent being the Higgs boson in 2012. The set of six quarks and six leptons has been extended as new particles are discovered, and the diagrams are used try to show these in some elegant configuration or model, for example Fig. 13.3.

---

[2] Niels Bohr was a cousin of Edgar Rubin, whose vase we saw in section 3.12 (Pind 2012).

For a physicist, it is an interesting but completely irrelevant coincidence that many of Bach's works are in sets of six, e.g. 6 Brandenburg Concertos, 6 Keyboard Partitas, 6 French Suites, 6 English Suites, 6 Violin Sonatas and Partitas, 6 Cello Suites, 6 Sonatas for Violin and Harpsichord, 6 Trio Sonatas for organ, 6 Schübler chorales, and in the current model of particle physics fermions come in 6 types of quarks (called up, down, strange, charm and bottom) and 6 types of leptons (called electron, muon, tau, electron neutrino, muon neutrino and tau neutrino). Sachs (1984) also points out that all Bach's instrumental works that are not cycles through the keys or variations are in sixes. Six has been regarded as the perfect number since antiquity as it is equal to the sum of its divisors: $1 \times 2 \times 3 = 1+2+3 = 6$. Bach's cousin (Walther 1732) gives a definition of this *numerus perfectus* in his musical dictionary of 1732, as (Tatlow 2015) explains, as a purely mathematical definition without putting it into a musical context. But as with Bach's works, there are other subatomic particles, bosons, that do not come in sixes.

Cosmologists make new discoveries as we build better telescopes to probe further into space and time and send satellites to ever more distant parts of the universe, or devise experiments to measure gravitational waves, but come up with bigger questions such as dark matter and dark energy. The "standard model" is an attempt to explain both subatomic and cosmological physics with a single elegant mathematical model.

In biology and life sciences, the diversity of life was first explained by Charles Darwin with evolution and natural selection in 1859. This was further refined in 1953 by James Watson and Francis Crick with the DNA double-helix model. The human genome (we saw the symbol set of the genetic code in Sect. 2.5) was decoded, in 2001 and now genetics is routinely used for medicine and forensics.

Models akin to these scientific models are also used in music. One example is tuning and temperament—the notes of the musical scale are defined by simple Pythagorean ratios, but tuning by the circle of fifths—tuning each note a perfect fifth above the previous note starting from C brings you back to a C, at first sight a very elegant model. However, this does not give a true octave but overshoots by a ratio of 129.74634:128 (Fig. 13.4—the lengths of the radials correspond to the frequencies of the notes), and so the ideal tuning has to be modified in some less elegant way, and there are many models for this from Werckmeister, Kirnberger and Suppig to the modern equal tempered.

Alternatively, instruments have been built to incorporate all the correct tones, for example the Orthotonophonium in Fig. 13.5, which for obvious reasons have never caught on.

If you listen to Bach's music, and especially if you play it, you find (as with any great music) that as you explore, more and more subtle aspects come to light—a phrase from one movement hidden in an inner voice of a preceding movement, a strange harmonic effect, an apparent change in tempo, the way a fugue or canon works out, etc.

It seems that with Bach's music, just as with the universe, we are confronted with something beautiful and interesting with inestimable depths of meaning which we feel the need to explain and understand to answer the questions: how did it

13.2 Models

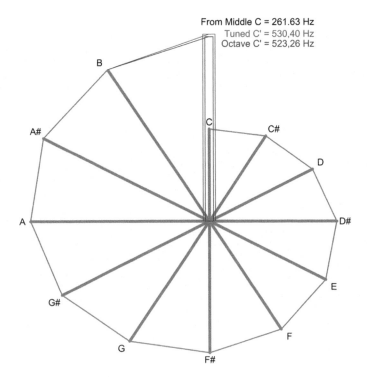

Fig. 13.4 Spiral of perfect fifths

come to be? and What does it mean? As with other branches of science, we make models, i.e. theoretical structures which explain the observed phenomena, and these are often mathematical. As discoveries or observations are made which break the current model, the model is refined or replaced with one which can explain the new discoveries. We feel that a correct model or mathematical explanation must be simple and elegant (the mathematician G.H. (Hardy 1967) said: "there is no permanent place in the world for ugly mathematics"), but as (Hossenfelder 2018) points out, there is no logical reason why this should be so. The representations of Vitruvian Man shown in Figs. 8.2 and 8.3 are models of the human body, the first being very simple and elegant, the second being more realistic and more complex.

To take a few more examples from music, Ruth Tatlow (Tatlow 2015) explains how Bach structured his works to achieve parallel layers of 1:1 and 1:2 proportions in the lengths, but to illustrate this has to occasionally make choices such as whether to count repeats or da capos or not.

Ulrich Siegele (Siegele 1997) explains how the works are structured to achieve a certain duration (see Sect. 10.9) and for the Goldberg Variations, shows how they can be divided into sections in various ways, but has to go across his groupings to achieve this. Having presented a structural model that divides the variations into three sets—character pieces, virtuoso pieces and canons, he finds that they do not quite fit

Fig. 13.5 Orthotonophonium in the Grassi Museum of musical instruments, University of Leipzig. Photograph © by the author with permission of the museum. Wiki Commons CC-BY-SA

into the columns, so needs to reallocate variations 2 and 28 as shown by the red and blue outlines in the left-hand part of Fig. 13.6. This does have a certain symmetry and forms a sort of interlocking between the sets, a refinement of the initial model. The right-hand side of Fig. 13.6 shows the simpler, more common model—there are two halves, the second starting with an overture, and every third variation is a canon on ascending notes until the last, which is a quodlibet instead of a canon at the tenth.

Magic squares are another type of model, for example a magic rectangle is obtained from the Well Tempered Clavier Book 1 by Henk Dieben (which we saw in Fig. 11.1) and (Mäser 2000) shows a magic square for Book 2 (Fig. 13.7) for which he needs to make some "corrections"—to obtain a square from 48 pieces there must be one empty cell (mathematically speaking a piece of length 0 bars) and the bars in the Prelude in D major from book 2 are counted without repeating the second half.

As the aphorism, attributed to the statistician George Box goes: All models are wrong, but some are useful.

Perhaps asking whether Bach intentionally created these structures, and if so, how he went about achieving them are the wrong questions—perhaps he just composed, and the numerical structures are simply a coincidental property and not part of his conscious design.

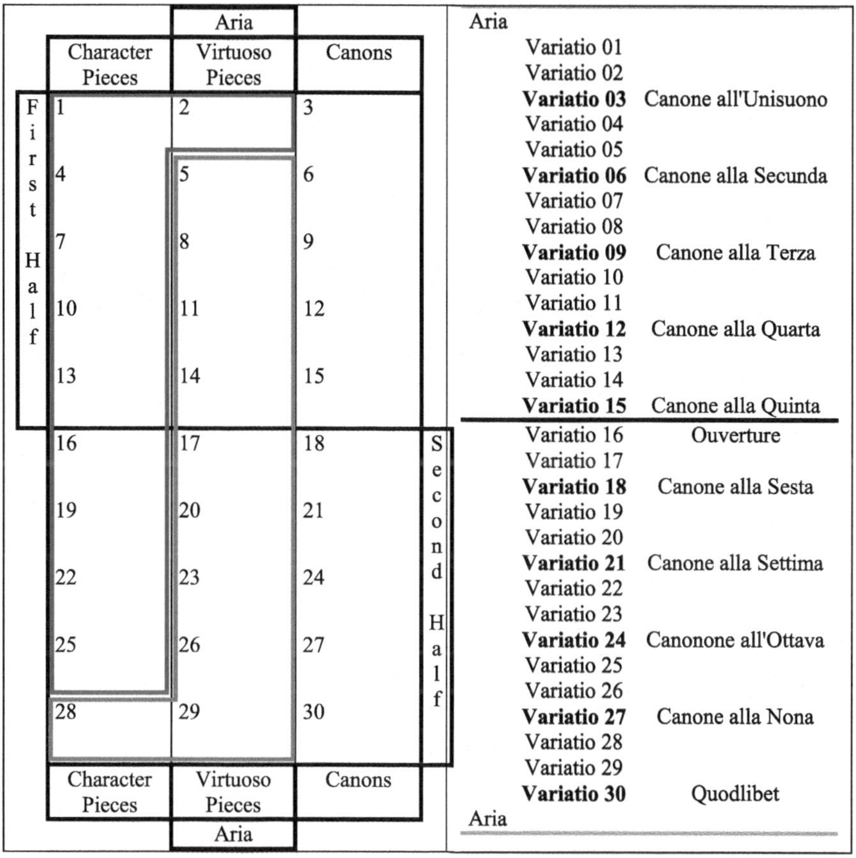

13.2 Models

	Aria			Aria	
	Character Pieces	Virtuoso Pieces	Canons	Variatio 01	
				Variatio 02	
First Half	1	2	3	**Variatio 03**	Canone all'Unisuono
				Variatio 04	
				Variatio 05	
	4	5	6	**Variatio 06**	Canone alla Secunda
				Variatio 07	
				Variatio 08	
	7	8	9	**Variatio 09**	Canone alla Terza
				Variatio 10	
	10	11	12	Variatio 11	
				Variatio 12	Canone alla Quarta
				Variatio 13	
	13	14	15	Variatio 14	
				Variatio 15	Canone alla Quinta
Second Half	16	17	18	Variatio 16	Ouverture
				Variatio 17	
				Variatio 18	Canone alla Sesta
	19	20	21	Variatio 19	
				Variatio 20	
	22	23	24	**Variatio 21**	Canone alla Settima
				Variatio 22	
				Variatio 23	
	25	26	27	**Variatio 24**	Canonone all'Ottava
				Variatio 25	
				Variatio 26	
	28	29	30	**Variatio 27**	Canone alla Nona
				Variatio 28	
				Variatio 29	
	Character Pieces	Virtuoso Pieces	Canons	**Variatio 30**	Quodlibet
		Aria		Aria	

Fig. 13.6 Structures of the Goldberg variations. Left: From Siegele (1997) with permission, modified and translated by the author. Right: By the author

46	61	28	216	29	28	96	Summe 504
84	50	77	83	100	75	35	504
64	50	100	34	70	43	143	504
72	174	70		56	99	33	504
101	27	140	21	93	72	50	504
66	71	43	72	84	83	85	504
71	71	46	78	72	104	62	504
504	504	504	504	504	504	504	Total 3528

Fig. 13.7 Magic rectangle from Mäser (2000). © Peter Lang used with permission

Fig. 13.8 Lull's concentric disc calculator. Lull (1274) Ars Magna https://archive.org/details/bub_gb_rG_yIN h8V1gC/page/n25/mode/2up Public Domain Mark 1.0

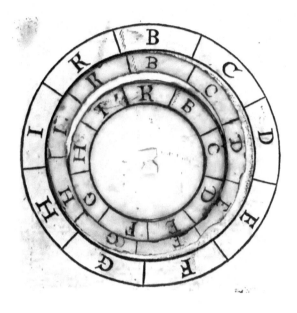

13.3 Computers

A further thread can be traced for the history of computers. Calculating machines could be said to start with the use of pebbles further refined by the abacus. The early users of combinatorics made concentric circular discs of paper which could be rotated to align various permutations of words or symbols, e.g. (Lull 1274) (Fig. 13.8 1308) for combining symbols representing concepts to examine their truth or Harsdörffer for permuting syllables to create all possible words as well as a circular slide rule for multiplication using combinations rather than logarithms (Fig. 13.9, 1651). Mizler (1739) invented one for calculating figured bass, but no image of this has survived. Mechanical calculators using wheels and gears were invented as early as 1623, and Leibniz presented one in 1673, which was the first that could multiply and divide as well as add and subtract—a modern reconstruction is shown in Fig. 13.10. Leibniz also developed binary notation in 1679, which we used for binary signatures in Sect. 8.2.6 and is the basis for all digital computers, information storage and communications. Further milestones are Charles Babbage's difference engine (1832) along with the first example of algorithms by Ada Lovelace, and the first electronic computers, first with relays, then valves,[3] transistors and finally integrated circuits. It is these that enable us to write such programs as the Proportional Parallelism Explorer and produce graphical representations of the results. It seems that the next step, or perhaps quantum leap, in computers will be the quantum computer, relying on quantum theory which deals in subatomic particles being in multiple states simultaneously with corresponding probabilities that determine what we actually see

---

[3] Also known as vacuum tubes.

Fig. 13.9 Harsdörffer's "Denckring" phrase calculator and circular slide rule. Schwenter (1692) Teil 2. http://digital.slub-dresden.de/id275480860 (Public Domain Mark 1.0)

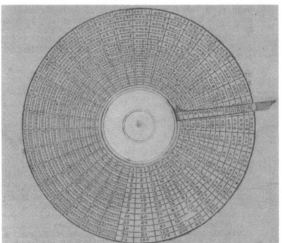

when we try to observe or measure them. We will get on to artificial intelligence in Sect. 13.4.

To connect with J. S. Bach again, one of the great computer scientists, Donald Knuth, author of the monumental "The Art of Computer Programming", has an organ in his house and composes music (Internet34).

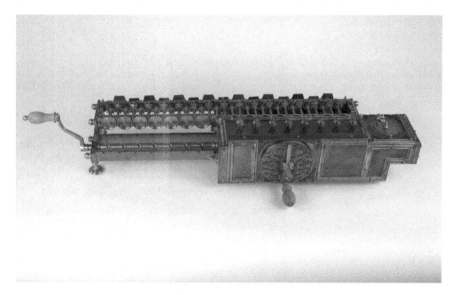

Fig. 13.10 Leibniz' mechanical calculator reproduction. © Arithmeum, Rheinische Friedrich-Wilhelms-Universität Bonn. Reproduced with permission

13.4 Artificial Intelligence

Computer-based techniques of artificial intelligence and machine learning started in the 1960s when Joseph (Weizenbaum 1966) wrote the ELIZA program. This was a surprisingly small program which simulates a Rogerian psychoanalyst and holds a conversation with the user via typed sentences, and even some experts could not tell that it was not a real analyst. Some advances have been made in this area, with computers beating world champions at chess and Go (a complex Japanese board game) as well as general knowledge quizzes. They are also being applied to music. They have been used, for example, to compose chorale harmonisations in the style of Bach which are able to fool experts [e.g. Hadjeres (2017), (Internet35), (Internet36)]. This worries some musicians, who think this will dehumanise the creative process thus making it meaningless, and as creativity is inherent to being human, this will devalue humanity.

But does art produced by artificial intelligence have value? We can differentiate between artistic value and the monetary value which may be achieved at auction. The monetary value should correlate with the artistic value but does not always, and it can vary with fashion and taste. The value of a work of art arises in a sort of spiral. Artists master the rules and techniques of their craft, in the end mastering the rules to such an extent that they will occasionally break them or invent new ones. They produce works which are excellent not only at the technical craft level, but also have something else which cannot easily be defined. Due to the way in which artists apply their craft, and due to their inner being, their works have a unique and identifiable style. When the artist becomes well known, their works become more expensive and earlier works also increase in value. The value of things usually increases in proportion to their rarity, and when an artist dies it is clear that no more works will be produced, and the value increases again.

Works of art can be forged, and forgeries can be very difficult to detect. A well forged copy of a painting or even a new work forged in the style of an artist may only be detectable by forensic analysis of the canvas and paint. People will initially appreciate a good forgery and value it just as much as an original work. But when it becomes clear that it is a forgery, it immediately loses its value.

Artworks produced by computers are essentially mass-produced imitations or forgeries. They may be interesting and pleasant, and they may fool people into thinking that they were created by the given artist or composer, but they have no value, because they are not by the artist and are not restricted in number.

This is not to say that humans do not engage in composing in the style of another composer. One example I remember is the harpsichordist George Malcolm's humorous fugue on the Sailors' Hornpipe in imitation of a Bach fugue, called "Bach before the Mast". Music is perhaps unique in that compositions are often modified by later composers, rearranged for other instruments or in another style. It was common practice in Bach's time to re-use their own work or adapt other composers' work, and it was a recognised teaching method to ask a student to add a bass line to a melody and then add a new melody to that bass line (see Tatlow 2015, p. 47). Bach arranged Italian orchestra concertos for organ in order to study them. Feruccio Busoni arranged some of Bach's works in the romantic style of his time. This "rearrangement" is less common in the visual arts, but (Walter 2011) shows examples of Arnulf Rainer's overpainting of works by old masters and there have been many "reinterpretations" of Leonardo da Vinci's Mona Lisa.

One might hope that generating works of art by computer in the style of a particular artist would give some insight into the artistic methods and inner processes used by the artist. This would be considered valuable. However, machine learning using complex simulations of neural networks cannot provide this as it does not produce a specific algorithm or method that can be accessed. This is one reason why many people fear artificial intelligence systems—although a complex traditional computer program can be very difficult to understand, it is still possible, and errors can (usually) be corrected. A self-learning AI system continually changes as it learns from the feedback telling it whether the previous result was good or bad, and it is not possible to determine exactly how it produced its result. Even if there were an accessible underlying algorithm it would not necessarily be the one used by the original artist. In the same way, the Proportional Parallelism Explorer program of Chaps. 8 and 9 (which is in no way intelligent, but uses a brute force approach) can find all 14,191 ways of obtaining a 1:1 proportion in WTC1, but it does not indicate which one Bach would have used, if indeed he intentionally used proportions. That can only be explored when the numerical evidence is supported by historical documents and methods, for example as in Tatlow (2015).

We can bring in some contemporary philosophy here—Leibniz distinguishes between logically necessary truths which can be demonstrated in a finite number of steps and contingent truths where an infinite number of steps is required and therefore cannot be calculated by humans (Butt 1997) p. 62–63). We could regard the combinations of pieces and proportional solutions as contingent, as the numbers, although not infinite, are often very large. Spinoza categorises knowledge with the first level as opinion or imagination, the second level as reason and the third level as intuition (Butt 1997, p. 69), giving an example of mathematical calculations for which the answer can be intuitive for small numbers but not for large numbers[4]: "Bach could apparently conceive of a musical idea and its extension in a single instant".

Going on to apply this to composition, Butt says: "Bach may well have thought in this manner when composing, determining which strategies followed by necessity from a particular theme or process (as determined quasi-mathematically, by the rules of harmony and counterpoint) and those which followed contingently, where composition involved discerning something of the predicates pertaining to the idea." A computer can find logically necessary truths, i.e. apply the rules of composition, but cannot find contingent truth, because this requires an infinite number of steps. (Whether a problem can be solved in a finite number of steps in a computer program is a fundamental area of research in computer science—look up "NP-Complete" and "Halting Problem" to find out more).

Ruth Tatlow mentioned in conversation that while working on her books she became quite adept at mentally calculating sums from numeric alphabets or proportions from lengths. It is therefore possible that Bach was more adept at these mental calculations than we might imagine today, with our reliance on pocket calculators and computers causing our mental arithmetic skills to be lost.

Automating the composition process is not new. Mersenne, Kirnberger, C.P.E. Bach, Mozart and others produced methods of composing a melody by using permutations and combinations of the notes of the scale or by using dice to select notes or prefabricated bars of music [(Klotz 2014), (Tatlow 2015)].

13.5 Quantz and Hi-Fi

While studying electronics in my youth, I was of course interested in high-fidelity sound systems. But since incremental improvements in sound quality come at an exponentially high price, at some point I asked myself: do I want to appreciate the quality of the sound system (wow! listen to that bass, oh! some harmonic distortion there), or do I want to listen to the music? I decided on the latter; albeit with a certain minimum standard of sound quality. Perhaps one needs to make a similar decision with Bach: do I want to understand the underlying structures and thought processes, or do I want to have the transcendental experience of just hearing or playing the music? Just as one cannot concentrate on the music from a hi-fi system if one is distracted by dissatisfaction with the sound quality, it is difficult to experience Bach's music as a listener or to immerse oneself in playing it and at the same time

---

[4]Butt does not think that this is a good example.

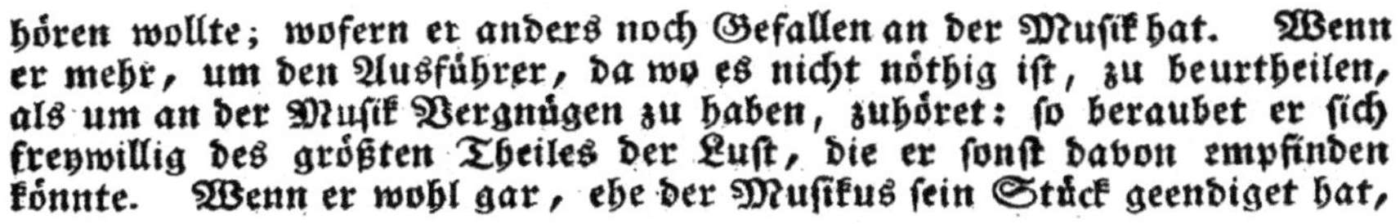

hören wollte; wofern er anders noch Gefallen an der Musik hat. Wenn er mehr, um den Ausführer, da wo es nicht nöthig ist, zu beurtheilen, als um an der Musik Vergnügen zu haben, zuhöret: so beraubet er sich freywillig des größten Theiles der Lust, die er sonst davon empfinden könnte. Wenn er wohl gar, ehe der Musikus sein Stück geendiget hat,

Fig. 13.11 Quantz (1752) extract from XVIII §7. With kind permission of Bärenreiter-Verlag Karl Vötterle GmbH & Co. KG

count the bars or entries of a fugue subject. Quantz (1752) had a similar idea in 1752, when of course there were no recordings and music could only be heard in live performances. He says in section XVIII §7 (Fig. 13.11, (Quantz 1752)): "If he [the critic] listens more to unnecessarily judge the performer than to enjoy the music, then he robs himself voluntarily of the largest part of the pleasure which he would otherwise experience".[5] (Tatlow 2015) also says: "the musicologist has to ... decide: either to give up any ambition to hear and think as Bach did, content to discover twenty-first-century resonances in his music; or to continue to attempt to hear as Bach did, and strive to understand the universe and music as he understood it."

13.6 Summary

There are a number of interesting links, both temporal and geographical, between the development of mathematics and statistics and J. S. Bach, but there is not enough evidence to show how much of this was known to Bach.

A path can be traced from the earliest attempts at calculators with rotating discs to obtain permutations, the use of permutations as a means of invention through to modern computers using artificial intelligence to compose music in a given style.

---

[5] Translation by the author.

Chapter 14
Conclusion

I set out to show how some areas of "hard" science, information theory, mathematics and computer science could be used in the humanities, specifically some aspects of musicology concerning numbers. It was not so much my intention to obtain specific results in musicology, as to introduce and demonstrate some methods which have not been widely used in these areas up to now.

For Gematria, I have provided a theoretical underpinning from information theory (Chap. 2) and given examples of how they could be encoded in music (Chap. 3).

I believe we have seen that some claims on numbers in Bach's music must be treated with scepticism (Chaps. 4 and 5) and that others can be rejected (Chap. 6), and in the latter, shown simple ways in which scores can be interpreted for computer analysis.

In the area of Proportional Parallelism, I have given an overview of some statistical methods (Chap. 7) and presented a rigorous terminology and a computerised tool for further exploration (Chaps. 8–10) which should make it easier to consider and discuss proportions in musical collections and works and to explore them much more easily than with manual techniques. I also related magic rectangles to proportions (Chap. 11); however, I have not managed to find a practical way of obtaining all the possible magic rectangles for a set of pieces.

I have hopefully made the reader aware of various psychological traps and pitfalls (Chap. 12), even though awareness of them does not always give protection against them.

Finally in Chap. 13, I have digressed a little to see how the techniques we have used relate to Bach and his time and looked out to some other sciences to see how they might relate to our subject.

I now leave it up to the real musicologists to use these methods and draw their own conclusions.

Many chapters have a "Notes for Researchers" section, and those are not repeated here. Additionally, I can offer the following more general notes:

Define terminology rigorously and use it consistently.

Declare all results and give enough information for others to reproduce them.

Use the simplest methods possible, both to avoid errors, and to enable replication by others.

Thoroughly test and validate computer programs and Excel sheets (Excel sheets are more error-prone than one might imagine).

Use graphics to make results more accessible.

Having a broad interest in various arts and sciences can give interesting insights. Research the history of techniques used and look out for interesting connections. Involve colleagues from other specialities.

Avoid cognitive biases as much as possible, do not ignore any inconvenient results.

Appendix A
More Parallel Proportion Results

A.1 Preamble

This chapter gives the detailed results for the works summarised in Chap. 10 apart from the Well Tempered Clavier Book 1(WTC1), which was covered in Chap. 9. It is not intended that this be read from beginning to end, but reducing it to the most interesting examples would be subject to biases and might give a misleading overall impression.

I generally only examine the 1:1 and 1:2 proportions as these are historically the most likely to be used. The key statistics for all the sets for these proportions were given in Table 10.22.

The works and collections are in alphabetical order. The naming convention for the collections and works with various ways of counting was given in Table 10.1.

The numbers of solutions and patterns are given in tables for each proportion. Note that the histogram tables give the numbers of samples that have no solutions as the first entry. The patterns being matched are only the symmetricalmirror imageor palindrome(Left=thgiR) and the translation(Left=Right) patterns. Proportions that are not possible are indicated by "-" entries.

A.2 Mass in B Minor BWV 232

The Mass in B minorwas composed by extending an existing Lutheran mass to create a Catholic mass for presentation to the August III, Elector of Saxony in 1733, who had converted to Catholicism so that he could become king of Poland.

Tatlow (2015) shows that there are two ways of counting the bars, either at the breve or at the semibreve, and she studies the rests and bar lines in the manuscripts and the orchestral parts to show how Bach used this. Tatlow (2015) also looks at the Lutheran version before addition of the Credo, Sanctus and Osanna. We can look at

Table A.1 Summary of simple proportions in Mass in B minor Lutheran version

	At the breve			At the semibreve		
	Solutions	Left=Right + Left=thgiR	Proportion count	Solutions	Left=Right + Left=thgiR	Proportion count
1:1	7	0	6	–	–	–
1:2	–	–	–	3	0 + 1	1

the five main sections (Kyrie, Gloria, Credo, Sanctus, Agnus Dei) or the 27 individual movements.

Lutheran Version

As this version only has two main sections, Kyrie and Gloria, we do not analyse these. Using the movements and counting at the breve, Tatlow (2015) shows a double 1:1 proportion and the program finds six and an additional one without the proportion in the count of pieces, but no symmetricalpatterns. See Table A.1.

Counting at the semibreve, only a 1:2 proportion is possible, for which there are three solutions with one mirror imagesymmetrical pattern and one with the double proportion in the number of pieces. Tatlow (2015) shows a symmetrical1:1 pattern of the three Kyrie movements against the first three movements of the Gloria, but this does not include all the movements. Other proportions are not possible.

There are no solutions with two layers.

The Monte Carlo simulations shows the difference between counting at the breve, which has 12 pieces, and counting at the semibreve, which only has nine pieces: the latter has the distinctive distribution for a small number of pieces with most of the samples yielding no solutions. The three real 1:2 solutions for semibreve lie in the 98th percentile of the simulation. The seven 1:1 real solutions for breve lie in the 73rd percentile which is lower than many other sets. These are shown in Fig. A.1.

Catholic Version Main Sections

There are only five main sections, Kyrie, Gloria, Credo, Sanctus and Agnus Dei in the Catholic mass, and these do not yield any proportions even where this would be possible—see Table A.2.

The Monte Carlo simulations also yield relatively few solutions, as we would expect (not shown).

Catholic Version Movements

Using the individual movements at the breve and at the semibreve (Table 13.5 in Tatlow 2015) gives the simpler proportions summarised in Table A.3.

Counting at the semibreve, Tatlow (2015) shows one 1:1 solution which includes all the 24 movements, and the program finds 10,807 such solutions with seven symmetricalpatterns. Tatlow (2015) shows other solutions, but none of these include

Appendix A: More Parallel Proportion Results

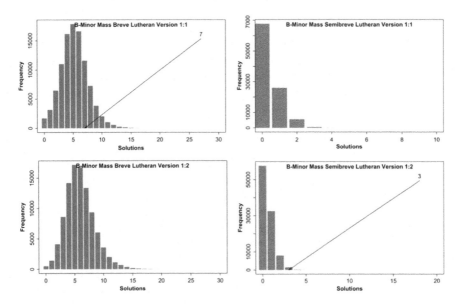

Fig. A.1 Monte Carlo simulations of Mass in B minor Lutheran version

Table A.2 Proportions in Mass in B minor main sections

	At the breve		At the semibreve	
	Solutions	Left=Right + Left=thgiR	Solutions	Left=Right + Left=thgiR
1:1	–	–	0	0
1:2	–	–	0	0

all the pieces. The program also gives thousands of solutions and several patterns for 1:2.

Of the 10,807 1:1 proportion solutions, 3731 have the doubleproportion in the number of pieces.

Second Layer

A second layer of 1:1 proportion gives 8838 stricttwo-layer solutions and 1710 with the additional layer of the proportion in the number of pieces.

The Monte Carlo simulationsfor layer 1 (Fig. A.2) show that the real number of solutions is in the 99th percentile for both proportions.

A.3 Brandenburg Concertos BWV 1046–1051

This is a collection of six concertos for various orchestral groupings which Bach presented to the Margrave of Brandenburg in 1721.

Table A.3 Simple proportions in Mass in B minor movements

	At the Breve			At the Semibreve			At the semibreve Two layers (Strict)		
	Solutions	Left=Right + Left=thgiR		Solutions	Left=Right + Left=thgiR	Proportion count	Solutions	Left=Right + Left=thgiR	Proportion count
1:1	–	–		10,807	2 + 5	3731	8838	2 + 0	1710
1:2	–	–		8,077	3 + 3	2153	–	–	–

Appendix A: More Parallel Proportion Results

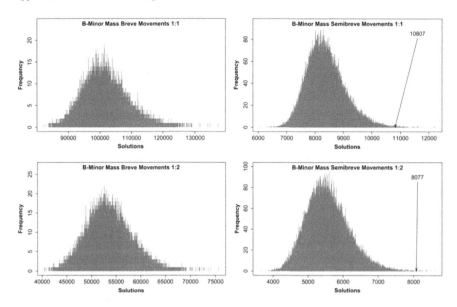

Fig. A.2 Monte Carlo simulations of Mass in B minor movements

The set cannot give any 1:2 proportions, and the six works do not produce any of the other proportions. The movements have plenty of solutions but few symmetricalpatterns. There are also solutions over two layers of proportion, but none having the parallel proportion in the number of pieces. See Table A.4.

The Monte Carlo simulations show that the real solutions are in the high 99th percentiles (Fig. A.3).

If we consider the sounding bars, we need to not only count all the repeats but also the reprises of the Menuetto in Concerto No. 1: Menuetto—Trio I—Menuetto—Polacca—Menuetto—Trio II—Menuetto. We must also decide whether the repeats of the Menuetto are to be played on every reprise. The CD recording by John Butt and the Dunedin Consort[1] does play all the repeats, and following this, we obtain 1:1 and 1:2 proportions, and there are far more solutions and more symmetricalpatterns—compare Table A.5 with Table A.4.

This is no surprise as we now have more pieces—25 instead of 22—and the Menuetto occurs four times.

The Monte Carlo simulations (Fig. A.4) show that the real number of solutions is in the 99.999th percentile for 1:1 and 1:2—the highest of any of the collections or works examined.

As in Sect. 10.12 about the Art of Fugue, we could speculate whether this gives us any indication as to what to do with the middle movement of the third concerto. Bach only wrote a single bar with two chords. Performers have applied various solutions to this puzzle, playing it as is, adding a harpsichord, violin or viola cadenza, or using

---

[1]Linn Records CKR 430, 2013.

Table A.4 Proportions in Brandenburg Concertos

	Works			Movements			Movements two layers(Strict)		
	Solutions	Left=Right + Left=thgiR		Solutions	Left=Right + Left=thgiR	Proportion count	Solutions	Left=Right + Left=thgiR	Proportion count
1:1	0	0		2200	1 + 2	0	1115	0	0
1:2	–	–		–	–	–	–	–	–

Appendix A: More Parallel Proportion Results 241

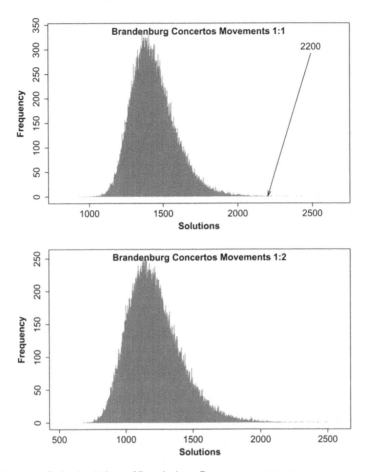

Fig. A.3 Monte Carlo simulations of Brandenburg Concertos movements

a movement from another work. If we regard the number of solutions as meaningful, we could test other options with cadenzas of different lengths to see what proportions would arise and how many solutions and patterns these would have.

A.4 Canonic Variations on "vom Himmel Hoch" BWV 769

These are five variations on the Christmas chorale in the form of canons which Bach submitted to Mizler's Corresponding Society of Musical Sciences in 1747.

There is only one 1:1 solution, the one given by Tatlow (2015) and this is not symmetrical (Table A.6). Tatlow also points out various proportions in the G-Values of the title page and incipits.

Table A.5 Proportions in Brandenburg Concertos with all Repeats

	Works		Movements			Movements two layers(Strict)		
	Solutions	Left=Right + Left=thgiR	Solutions	Left=Right + Left=thgiR	Proportion count	Solutions	Left=Right + Left=thgiR	Proportion count
1:1	0	0	17,185	1 + 2	–	12,185	0	–
1:2	0	0	17,422	6 + 4	–	11,517	0	–

Appendix A: More Parallel Proportion Results

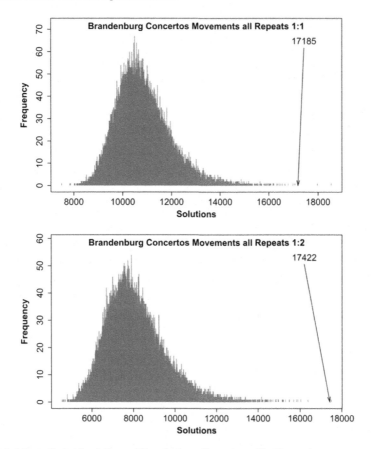

Fig. A.4 Monte Carlo simulations of Brandenburg Concertos with all repeats

Table A.6 Proportions in Canonic Variations

	Canonic variations		
	Solutions	Left=Right+ Left=thgiR	Proportion count
1:1	1	0	–
1:2	–	–	

The Monte Carlo simulations show that more solutions would have been possible, although as we would expect for only five pieces, not as many as in larger collections.

Table A.7 Monte Carlo simulations of Canonic Variations

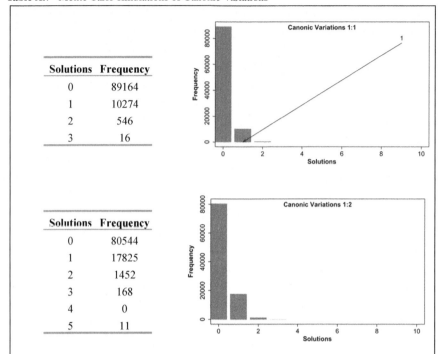

Solutions	Frequency
0	89164
1	10274
2	546
3	16

Solutions	Frequency
0	80544
1	17825
2	1452
3	168
4	0
5	11

A.5 Cello Suites BWV 1007–1012

There is no original manuscript for these, only different copies including one by Bach's second wife Anna Magdalena. They were probably composed in Bach's Weimaror Köthen period, before 1720. They were forgotten until the young Pablo Casals discovered an edition in a second-hand shop in 1898. Anyone performing them must make many decisions when choosing which copy to follow. This will be facilitated by new media such as that suggested by Szabó (2016) and demonstrated in prototype at the Bach Network Dialogue Meeting 2019.

Tatlow (2015) shows various ways of counting the bars in the six cello suitesusing Anna Magdalena Bach's copy.

Omitting the repeatsand da capos for a total of 1971 bars, 1:1 is not possible and there are no 1:2 solutions. Taking the 42 movements, there are 2,263,016,866 solutions and 2694 patterns(1665 "Left=Right" and 1029 "Left=thgiR") for the 1:2 proportion (Table A.8). (The program took over 2½ days to run—each additional movement doubles the number of possible combinations—see Sects. 8.2.3 and B.23.)

Omitting the repeats but including da capos (except for the da capo of Bourée 1 in BWV 1010) and including the extra 4 bars dal segno in the Gavotte of BWV 1012 as indicated in Tatlow (2015) Table 5.12 giving 2115 bars, the 1:2 proportion given

Appendix A: More Parallel Proportion Results 245

there is the only one, with suites 1, 2, 3 and 5 against 4 and 6—a double 2:1 with the number of pieces. 1:1 is not possible.

For the version using Anna Magdalena Bach's copy including all repeatsand da capos to give 4000 bars, the round number typical of Bach's finished collections, there are no 1:1 solutions for the six works, and 1:2 is not possible. The 48 movements are too many to run the program to completion. As one might expect, there are large numbers of solutions and patterns (see Table A.9).

Running a Monte Carlo simulation of 100,000 or even 1000 samples when each sample potentially takes several years is even less feasible.

A.6 Clavierübung I—Partitas BWV 825–830

The six partitas for keyboard, meaning harpsichordor clavichord, were published by Bach as a set in 1731, but written and published separately from 1726. As an amateur musician, I always think the title "Keyboard Exercise No. 1" is rather misleading, as they are not easy to play!

The Works

At the level of the six works, there are no 1:1 or 1:2 proportions.

The Monte Carlo simulation shows that there could have been solutions with other lengths, and there is a 0.2% chance of finding a solution (an average of 0.063 solutions over 31 possible combinations).

The Movements

The 41 movements took over seven days to run the program to completion for 1:1. There are over 2 billion 1:1 solutions and thousands of symmetricalpatterns—see Table A.11.

Running a Monte Carlo simulation for these is prohibitive, as each sample will potentially take a week to run.

A.7 Clavierübung II—Italian Concerto BWV 831 and French Overture BWV 971

The second "Keyboard Exercise" followed on from the first and was published in 1735.

The two works do not have any proportion between the 451 and 491 bars.

The 14 movements, which is below the threshold for being sure of finding solutions, have no solutions for 1:1, but there are 27 solutions for 1:2 with no patterns and no parallel proportion in the counts of pieces (see Table A.12).

Table A.8 Proportions in six Cello Suites

	No repeats or da Capos (1971 bars)				No repeats with da Capos (2115 bars)			
	Works		Movements		Works		Movements	
	Solutions	Left=Right + Left=thgiR	Solutions	Left=Right + Left=thgiR	Solutions	Left=Right + Left=thgiR	Solutions	Left=Right + Left=thgiR
1:1	–	–	–	–	–	–	–	–
1:2	0	0	2,263,016,866	1665 + 1029	1	0		

Appendix A: More Parallel Proportion Results

Table A.9 Partial results for A. M. Bach's copy of Cello Suites

Proportion	No. tried	% tried	Time taken	No. solutions	No. patterns
1:1	2,009,978,736,390	1.43%	17 days	9,369,844,255	1126
1:2	–	–	–	–	–

Table A.10 Monte Carlo simulation of Clavierübung I works 1:1

Solutions	Frequency
0	93863
1	5961
2	174
3	0
4	2

Table A.11 Proportions in Clavierübung I movements

	CÜ1 movements	
	Solutions	Left=Right + Left=thgiR
1:1	2,235,988,612	1628 + 1690
1:2	854,821,792	1059 + 1339

Table A.12 Proportions in Clavierübung II movements

	CÜ II movements		
	Solutions	Left=Right + Left=thgiR	Proportion count
1:1	0	0	0
1:2	27	0	–

The Monte Carlo simulations (Fig. A.5) show that there could have been solutions for 1:1. For 1:2, the 27 solutions found are in the 99th percentile.

If we count all the repeatsand da capos of the French Overture to give the length played, the total of 1497 bars in 15 pieces can only give a 1:2 proportion and I do not consider it further.

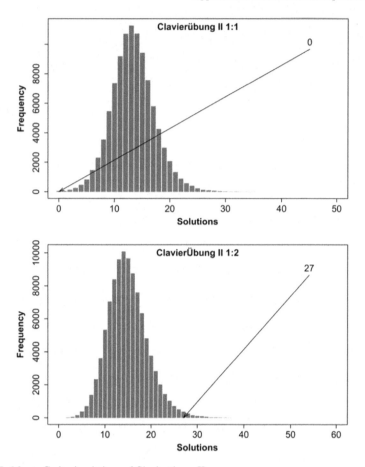

Fig. A.5 Monte Carlo simulations of Clavierübung II

A.8 Clavierübung III—Organ Mass BWV 669–689, 552, 802–805

The third "Keyboard Exercise" published in 1739 is a diverse set of chorale and mass settings for the organ, although not all pieces use the pedals, and some could be played on the harpsichord.

For Clavierübung III, Tatlow (2015) gives two alternative bar counts, the second counting repeats and including a time signature change as a bar boundary.

As we would expect for a collection of 27 pieces with a wide range of lengths between 15 and 205 bars, there are many proportional solutions and a number of symmetricalpatterns. For the 1:2 proportion, there are 79,524 solutions of which 16,233 have a parallel proportion in the number of pieces—see Table A.13.

Appendix A: More Parallel Proportion Results

Table A.13 Proportions in Clavierübung III

	CÜ 3			CÜ 3 alternative		
	Solutions	Left=Right + Left=thgiR	Proportion count	Solutions	Left=Right + Left=thgiR	Proportion count
1:1	124 760	11 + 9	–	–	–	–
1:2	–	–	–	79,524	13 + 9	16,233

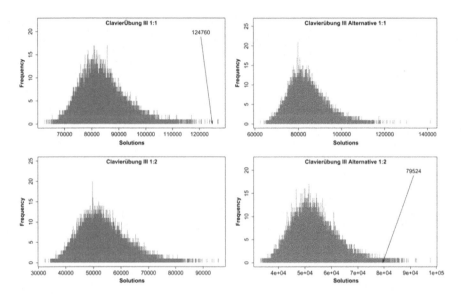

Fig. A.6 Monte Carlo simulations of Clavierübung III

The Monte Carlo simulations show that the real collection is in the 99th percentile of random collections—see Fig. A.6.

A.9 English Suites BWV 806–811

This is Bach's earliest set of suites written in the 1720s and not published by him. The title "English" does not mean that they are in the English style but possibly denote that they were written for an English patron.

We use the lengths given in Tatlow (2015) taken from Kayser's copy.

The Works

The six works do not have any solutions for 1:1 or 1:2 proportions. The Monte Carlo simulations (Table A.14) show that solutions would be possible.

Table A.14 Monte Carlo simulation of English Suites works

Solutions	Frequency
0	90850
1	8656
2	488
3	3
4	3

Solutions	Frequency
0	85657
1	13519
2	644
3	173
4	0
5	7

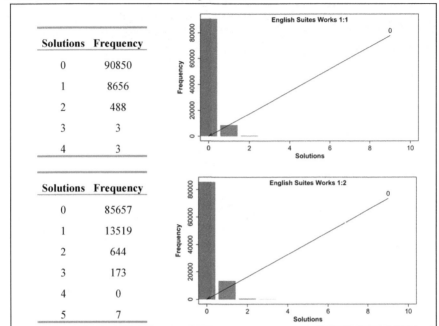

The Movements

The 42 movements are too many to run the search for solutions to completion, but a thirteen-hour run showed several hundred thousand 1:1 solutions having only tried about 1.6% of the possible combinations. For 1:2 a sixteen-hour run gave over half a million solutions having tried about 0.6% of the combinations. This is not surprising with the large number of pieces (as we saw in 10.2) and 30 of the 42 movements have lengths divisible by 8.

I did not search for patterns in these runs, but sorted the pieces by descending length to optimise the speed.

A Monte Carlo simulation is not feasible.

A.10 French Suites BWV 812–817

The French Suites are not particularly French in style and it is not clear where the name came from. They were written from 1722, initially into the notebook of his second wife Anna Magdalena and were probably combined into the set in 1725.

I use the lengths given in Tatlow (2015) taken from Altnickol's copy.

Appendix A: More Parallel Proportion Results

Table A.15 Monte Carlo simulations of French Suites works

Solutions	Frequency
0	91522
1	8108
2	363
3	3
4	4

Solutions	Frequency
0	87014
1	12317
2	528
3	135
4	0
5	6

Table A.16 Proportions in the French Suites movements

		French suites		
	Solutions	Left=Right + Left=thgiR	Proportion count
1:1	103,600,369	576 + 618	–
1:2	44,821,980	512 + 551	–

The Works

The six works as a set do not have any solutions for 1:1 or 1:2 proportions. The Monte Carlo simulations show that solutions would be possible (Table A.15).

The Movements

We obtain large numbers of solutions for the 35 movements with the simple proportions, and neither 1:1 nor 1:2 have a layer of proportion in the numbers of pieces (Table A.16).

One mirror image 1:1 pattern stands out, shown in Table A.17, with the first and last suites in the solution and the third and fourth in the complement. The axis of symmetry is at row BWV814/6. (The proportion is not present in the number of pieces.)

The Monte Carlo simulations are shown in Fig. A.7. The collection gives large

Table A.17 Pattern in French Suites 1:1

Index	1252400S	1252400C
Layer	1	1
Length	690	690
Count	18	17
Pattern	Left=thgiR	Left=thgiR
BWV812/1	24	
BWV812/2	24	
BWV812/3	24	
BWV812/4	48	
BWV812/5	48	
BWV812/6	28	
BWV813/1		18
BWV813/2	54	
BWV813/3	24	
BWV813/4	32	
BWV813/5		16
BWV813/6		84
BWV814/1		24
BWV814/2		28
BWV814/3		24
BWV814/4		60
BWV814/5		72
BWV814/6		32
BWV814/7		68
BWV815/1		20
BWV815/2		36
BWV815/3		24
BWV815/4		44
BWV815/5		60
BWV816/1		24
BWV816/2	32	
BWV816/3	40	
BWV816/4	70	
BWV816/5		56
BWV817/1	28	
BWV817/2	32	
BWV817/3	24	

(continued)

Appendix A: More Parallel Proportion Results

Table A.17 (continued)

BWV817/4	24	
BWV817/5	86	
BWV817/6	48	

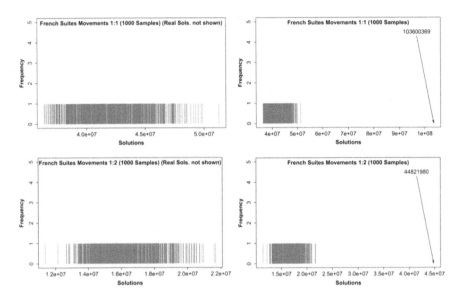

Fig. A.7 Monte Carlo simulations of French Suites movements

numbers of solutions, so they are shown on the horizontal axis in exponential notation, e.g. 4.0e+07 means 4.0×10^7, a four with seven zeros or 40,000,000. As we could only run 1000 samples, we do not see all possible numbers of solutions, and none occurs more than once. The real number of solutions is well beyond the right margin of the histograms, and this is shown on the right-hand side of Fig. A.7 scaled to make the real number of solutions visible. For example, for the 1:1 proportion, the real number of solutions is 103,600,369 which would be represented as 1.03e+08—well off the scale of the top left histogram and shown in relation to this on the top right.

Table A.18 Proportions in Goldberg Variations

	At the breve	
	Solutions	Left=Right + Left=thgiR
1:1	156,306,160	2100 + 1716
1:2	58,988,215	1078 + 1463
2:3	172,844,425	1870 + 2497

A.11 Goldberg Variations BWV 988

The name comes from Johann Gottlieb Goldbergwho was the harpsichordist of Count Keyserlingk, an influential diplomat in Dresden, who probably supported Bach in his request to the Elector of Saxony to be given the title of court composer when he submitted the Mass in B minor. The Variations were written for Goldberg to play to the Count when suffering from insomnia and published in 1741. They were titled "Clavierübung" but the number IV, following on from the previous three "Keyboard Exercises" is a modern term.

The work gives very large numbers of possible proportions because the thirty-two pieces are mostly the same length of 32 bars, with five of 16 bars and one of 48 bars (counting as given in Tatlow (2015) Table 7.6 with variation 22 counted alla breve).

There are 156,306,160 1:1 solutions which divided by $(2^{32}/2) - 1$ gives a probability of 7.3%—as expected, relatively high compared to other works and collections.

This work can also be used to verify that the explorer program is finding all the solutions (this was of course already verified against the manually worked example in Table 8.6). We can find the number of 1:1 solutions manually as follows.

The 32 pieces are made up of:

- 26 pieces of 32 bars each,
- 5 pieces of 16 bars each,
- 1 piece of 48 bars.

These can be combined in the following ways to give the 1:1 target value of 480 (half of the total of 960 bars):

48	16	32	48	16	32	Total combinations
Omitting the 48 and	omitting the 16 s and	using any 15 of the 32 s	0	0	$^{26}C_{15}$	7,726,160
	using any 2 of the 16 s and	using any 14 of the 32 s	0	$^{5}C_{2}$	$^{26}C_{14}$	96,577,000
	using any 4 of the 16 s and	using any 13 of the 32 s	0	$^{5}C_{4}$	$^{26}C_{13}$	52,003,000

(continued)

(continued)

48	16	32	48	16	32	Total combinations
including the 48 and	using any 1 of the 16 s and	using any 13 of the 32 s	1	$^5C_1$	$^{26}C_{13}$	52,003,000
	using any 3 of the 16 s and	using any 12 of the 32 s	1	$^5C_3$	$^{26}C_{12}$	96,577,000
	using any 5 of the 16 s and	using any 11 of the 32 s	1	$^5C_5$	$^{26}C_{11}$	7,726,160
					Sum	312,612,320
		/2 as the sum includes the complements				156,306,160

See the box in Sect. 8.2.3 for the $^5C_2$ notation of combinations.

A Monte Carlo simulation will not tell us very much, because there are only three different lengths in the real work and two of them (48 and 32) are multiples of the third (16). The results are shown in Fig. A.8. There are many different numbers of solutions, each occurring only once. The 1000 samples are not nearly enough to include the real set. This can be seen on the right-hand histograms of Fig. A.8, which shows the same data as the corresponding left-hand histogram scaled to include the real set. The real solutions are off the right-hand side of the horizontal axis.

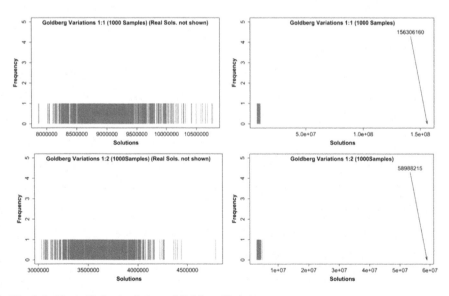

Fig. A.8 Monte Carlo simulations of Goldberg Variations

Table A.19 Proportions in Great 15 Organ Preludes

	Solutions	Left=Right + Left=thgiR	Proportion count
1:1	33	0	–
1:2	39	1 + 0	12

A.12 Great Fifteen Organ Preludes BWV 651–665

These were written in 1746–1747 and there is some doubt as to the intended number as some were added later by other copyists, going up to BWV 668 to give the eighteen more usually known. The "Great" in the title originates with Mendelssohn when he published them in 1846, part of the so-called Bach revival started by Mendelssohn's performance of an abridged version of the St. Matthew Passion in 1829. (I say "so-called revival" because all the great composerssuch as Mozart, Beethoven, Haydn possessed and studied copies of Bach's works, they were just not performed in public.)

We follow Tatlow (2015) in looking at the original fifteen pieces, and she shows one 1:1 and one 1:2 solution. The program finds thirty-three 1:1 and thirty-nine 1:2 solutions. Of these, twelve of the 1:2 solutions have a parallel proportion in the count of pieces and one of those has a symmetry. See Table A.19.

Tatlow (2015) also presents a second layer dividing the complement of the 1:1 solution into a further 1:1, and a second layer dividing the 1:1 solution into a 1:2. Of the thirty-nine 1:1 solutions, sixteen have a second layer, but none are strict. The 1:2 proportion cannot have a second layer as the 400:800 bars are not divisible by 3.

The Monte Carlo simulations (Fig. A.9) show that the real collection is not quite in the highest percentiles as many of the others.

A.13 Inventions and Sinfonias BWV 772–801

As Separate Works

The 15 Inventions alone and the 15 Sinfonias alone can only have 1:1 solutions, and only the Sinfonias have any symmetricalpatterns, as shown in Table A.20.

As a Collection

Taken together, the 30 pieces have 1:1 and 1:2 proportions as shown in Table A.21, with many double proportions in the number of pieces.

The Monte Carlo simulations are shown in Fig. A.10. For 1:1, the real collection is in the 99.99th percentile of the simulation, and for 1:2, the real collection has more solutions than were found in the simulation even though we have 100,000 samples.

Appendix A: More Parallel Proportion Results 257

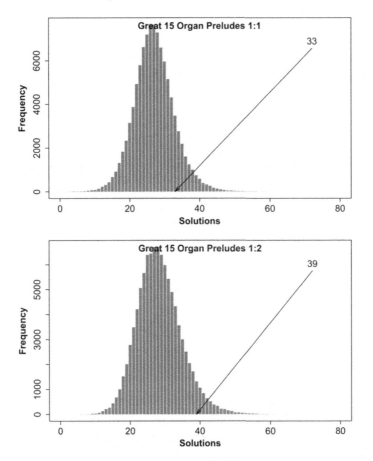

Fig. A.9 Monte Carlo simulations of Great 15 Organ Preludes

Table A.20 Inventions and Sinfonias as separate works

	Inventions			Sinfonias		
	Solutions	Left=Right + Left=thgiR	Proportion count	Solutions	Left=Right + Left=thgiR	Proportion count
1:1	94	0	–	79	2 + 0	–
1:2	–	–	–	–	–	–

Table A.21 Inventions and Sinfonias as one collection

	Inventions and Sinfonias		
	Solutions	Left=Right + Left=thgiR	Proportion count
1:1	2,072,960	43 + 45	766,567
1:2	1,040,660	40 + 45	312,771

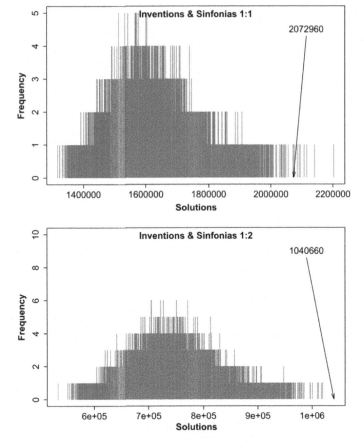

Fig. A.10 Monte Carlo simulations of Inventions and Sinfonias

A.14 Musical Offering BWV 1079

The Musical Offeringstems from Bach's visit to the Potsdam court of Frederick the Great in 1747 where Bach's son Carl Philipp Emanuel was keyboard player. Frederick was a keen amateur flautist, taught by Quantz. Bach tried out the new fortepianos and Frederick played a theme and asked Bach to improvise a fugue on the theme. On his return to Leipzig, Bach wrote out several fugues (ricercars), canons and a sonata based on the king's theme and sent them to him as the Musical Offering. It was also published for general sale.

As this is a work for presentation to the king, we would expect Bach to have taken extra care in its construction. Tatlow (2015) treats this thoroughly, including the G-Values of the cover page and titles of the pieces.

The 1100 bars of the ten pieces in the original print could give 1:1 or 2:3 proportions, but 1:1 yields none and 2:3 only the two solutions given by Tatlow (2015) (see

Appendix A: More Parallel Proportion Results

Table A.22 Proportions in Musical Offering

	Musical offering	
	Solutions	Left=Right + Left=thgiR
1:1	0	0
1:2	–	–
2:3	2	0

Table A.23 Architectural proportion in Musical Offering 2:3

Title	Section	2:	3
Ricercar à 3	B	185	
Canon Cancrizans	B	5	
Sonata	C		440
Canon perpetuus (Flute)	C		32
Canon perpetuus (Violin)	C		32
Canon perpetuus (Continuo)	C		32
Canones Diversi	D		124
Ricercar à 6	E	206	
Canon à 2	E	16	
Canon à 4	E	28	
		440	660

Table A.22). With only ten pieces, we are well below the threshold for being certain to find 1:1 proportions (see Sect. 10.2). (1:3 is also possible and yields 2 solutions).

One of the solutions for the 2:3 proportion corresponds to an architectural aspect, with sections B and E against C and D, although these are a division by modern scholars and not by Bach. This is given in Tatlow (2015) and shown in Table A.23. Section A is the title page.

The Monte Carlo simulations show that all the basic proportions would have been possible, and the solutions for 2:3 are in the 77th percentile (Fig. A.11).

A.15 Schübler Chorales BWV 645–650

These six chorales were printed around 1748 by the engraver Schübler.
Tatlow (2015) gives three alternative counts for these

- 255 bars with no repeatsor da capos—this can have 1:1 or 2:3 proportions
- 280 bars plus da capos but without repeats giving eight pieces—this can have 1:1 or 2:3 proportions
- 317 bars counting both da caposand repeats in six pieces—this cannot have any proportions as 317 is a prime number.

Solutions	Frequency
0	60752
1	30214
2	7773
3	1083
4	157
5	16
6	4
7	1

Solutions	Frequency
0	49587
1	35486
2	11480
3	2756
4	501
5	146
6	33
7	3
8	4
9	4

Solutions	Frequency
0	42213
1	35393
2	16597
3	4533
4	1052
5	156
6	42
7	9
8	5

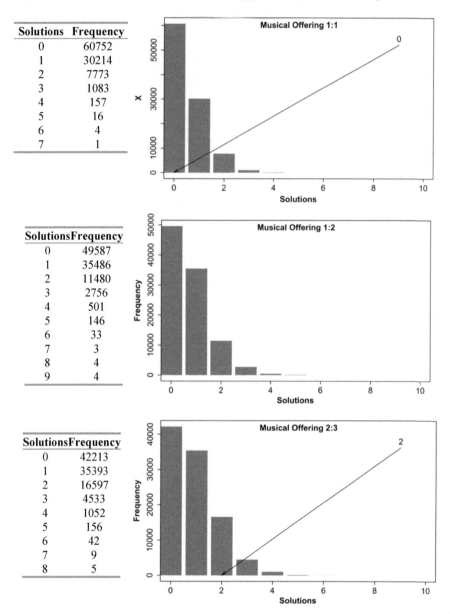

Fig. A.11 Monte Carlo simulations of Musical Offering

Table A.24 Proportions in Schübler Chorales

	Without repeats or da capos (255 bars, 6 pieces)		Repeats plus da capos separate (280 bars, 8 pieces)		Repeats with da capos included (280 bars, 6 pieces)	
	Solutions	Left=Right + Left=thgiR	Solutions	Left=Right + Left=thgiR	Solutions	Left=Right + Left=thgiR
1:1	–	–	1	0	0	0
1:2	0	0	–	–	–	–

Table A.25 Monte Carlo simulations of Schübler Chorales without Repeats or Da Capos

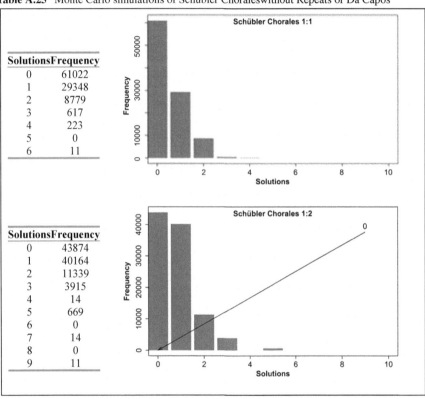

Solutions	Frequency
0	61022
1	29348
2	8779
3	617
4	223
5	0
6	11

Solutions	Frequency
0	43874
1	40164
2	11339
3	3915
4	14
5	669
6	0
7	14
8	0
9	11

The second alternative seems somewhat artificial as it takes the da capo bars as separate pieces giving eight instead of six pieces, and this only has the one solution given by Tatlow (2015). If we count the da capos within the six pieces, we obtain no 1:1 proportions (see Table A.24).

With only six pieces, we are not surprised to find so few solutions and that the eight pieces have slightly more (see Sect. 10.2).

The Monte Carlo simulations (Table A.25) show that the version with eight pieces

Table A.26 Monte Carlo simulations of Schübler Chorales plus Da Capos

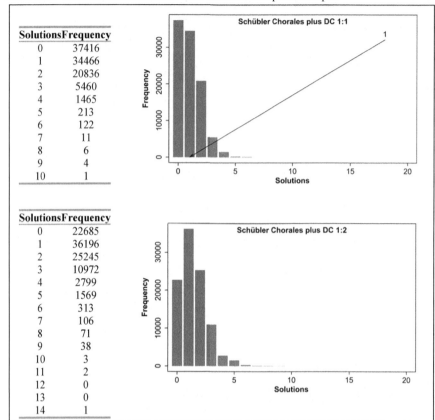

Solutions	Frequency
0	37416
1	34466
2	20836
3	5460
4	1465
5	213
6	122
7	11
8	6
9	4
10	1

Solutions	Frequency
0	22685
1	36196
2	25245
3	10972
4	2799
5	1569
6	313
7	106
8	71
9	38
10	3
11	2
12	0
13	0
14	1

gives better looking distributions with more solutions, but this is purely due to having more pieces. More solutions would have been possible. For the 1:1 proportions plus da capos, the real number of solutions is only in the 37th percentile of the simulation, which is lower than for most other collections.

A.16 Sei Soli for Violin BWV 1001–1006

This set of three sonatas and three partitas for violin solo was composedin Köthen around 1720. We already came across the six works in Sect. 10.4 and the Ciaccona from Partita no. 2 in Chap. 6.

Appendix A: More Parallel Proportion Results 263

Table A.27 Monte Carlo simulations of Schübler Chorales including Da Capos

Solutions	Frequency
0	72673
1	22953
2	4158
3	152
4	62
5	0
6	2

Solutions	Frequency
0	58996
1	33567
2	5450
3	1789
4	5
5	191
6	0
7	1
8	0
9	1

The Six Works

For the six works (Table 5.1 in Tatlow (2015)), there is no combination of the six that gives a 1:1 proportion in the lengths (but there is the 3:3 Sonatas:Partitas).

There is only the one possible 1:2 combination given by Tatlow, which also has the parallel proportion in the number of pieces 2:4. The Monte Carlo simulation (Table A.28) shows that the double 1:2 proportion in bars and count of pieces has almost as many solutions as the proportion in bars alone—the difference is not visible on the histograms. The average number of double 1:2 proportions is 0.02866, which dividing by the 64 possible combinations gives a probability of 0.045%. The real collection is in the 97th percentile of the random samples.

The 1:1 proportion does not occur for this collection, but we can still run the Monte Carlo simulation, and indeed, it would be possible to compose lengths in this proportion (see Table A.29). The average number of solutions for the double 1:1 proportion per sample is 0.029. There are $2^6/2 - 1 = 31$ combinations for the six pieces and a 1:1 proportion. Therefore, the probability of finding a double 1:1 proportion is $0.029/31 = 0.00092$ or 0.092%, about twice that of 1:2.

Table A.28 Monte Carlo simulation of Sei Soli 1:2

Solutions	Bars	Pieces
0	95203	95609
1	4710	4319
2	73	58
3	14	14

Table A.29 Monte Carlo simulation of Sei Soli 1:1

Solutions	Bars	Pieces
0	96959	97179
1	2992	2777
2	48	43
3	1	1

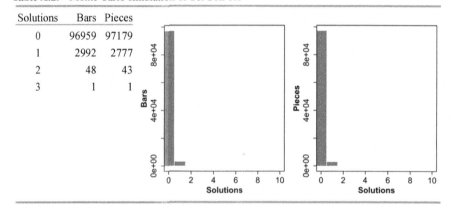

Sonatas and Partitas Individually

Taking the movements within the individual works, although they all have an even number of bars and so could give 1:1 proportions, as Tatlow (2015) shows, only BWV 1001, 1002 and 1006 actually have a 1:1 proportion.

For BWV 1001, the solution for 1:1 given by Tatlow (2015) is the only one.

BWV 1002 has eight solution pairs for the 1:1 proportion in the numbers of bars (facilitated by the fact that the Doubles each have the same number of bars as their predecessor movements). Of these, all have the double proportion in the number of pieces and three have symmetricalpatterns, two "Left=Right"s (one of which is Tatlow's) and two "Left=thgiR"s (one of which is also Left = Right). There are nine solution pairs for 1:2 but no symmetries. None of them reflect the proportion in the number of pieces.

There are no two-layer solutions in any of the works.

The results are summarised in Table A.30.

Appendix A: More Parallel Proportion Results 265

Table A.30 Proportions in BWV1001, BWV 1002 and BWV 1006

	BWV 1001			BWV 1002				BWV 1006	
	Solutions	Left=Right + Left=thgiR	Proportional count	Solutions	Left=Right + Left=thgiR	Proportional count	Solutions	Left=Right + Left=thgiR	Proportional count
1:1	1	0	0	8	2 + 2	8	2	0	–
2 Layers	0	0	0	0	0	0	0	0	–
1:2	–	–	–	9	0	–	–	–	–
2 layers	–	–	–	–	–	–	–	–	–

The Movements

We now consider the individual movements of the Sonatas and Partitas as a set with a total of 32 movements as the pieces in the set.

1:1 Proportion

For a 1:1 proportion, there are 2,602,210 solutions with 39 exhibiting a mirrored left/right symmetry and another 39 with the same sequence in the left and right halves.

The Monte Carlo simulation took 12 h for 1000 samples, so doing 100,000 was not practical as it would have taken about 50 days. The length histogram (Fig. A.12) therefore shows a less uniform distribution than we would like, but it can still be useful.

The results (Fig. A.13) show an average of 1,428,975 solutions with a minimum of 1,132,789 and a maximum of 1,853,843 solutions. The real collection has 2,602,210 solutions, more than any of the simulated samples, and nearly twice as many as the average of the thousand samples. Note also that because there are so many solutions (over 1.8 million in this simulation), each number of solutions only occurs once if at all, or at the most twice, giving a distribution which looks like a uniform rather than a normal distribution.

With a more powerful computer with six processor cores, it was possible to run several Monte Carlo simulations in parallel by starting the explorer program five times (leaving one processor for me to work with!). Running five simulations of 1000 samples in parallel also took about 12 h, and the results were combined using Excel. As expected, the lengths histogram (Fig. A.14) shows a better uniform distribution. The range of the numbers of solutions is also increased, but still does not achieve that of the real collection (Fig. A.15).

1:1 with Layers

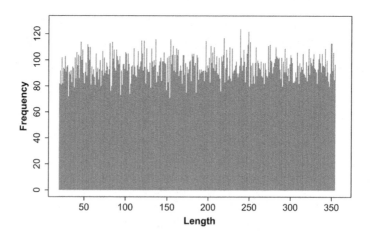

Fig. A.12 Sei Soli movements 1:1—lengths histogram for 1000 samples

Appendix A: More Parallel Proportion Results

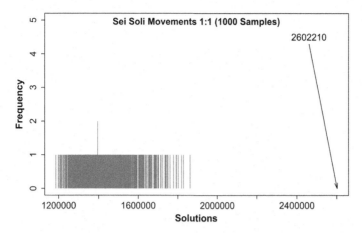

Fig. A.13 Sei Soli movements 1:1—results histogram for 1000 samples

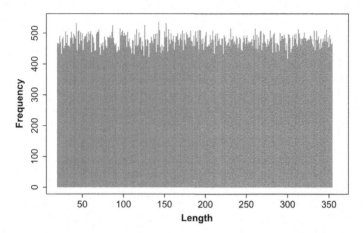

Fig. A.14 Sei Soli movements 1:1—lengths histogram for 5000 samples

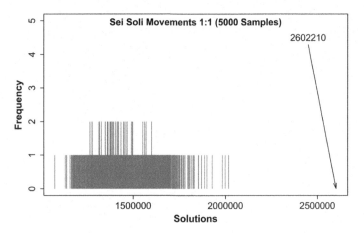

Fig. A.15 Sei Soli movements 1:1—results histogram for 5000 samples

For a second layer, dividing the 1:1 solutions of 1200:1200 bars further into 600:600 bars, the program finds 2,501,937 strict solutions for two layers, so 96% of the solutions found above can be extended to the second layer in multiple ways.

The first strict solution found by the program is shown in Table A.31. The layer 1 solution 1S divides into 1S_1S and 1S_1C at layer 2, and the complement 1C divides into 1C_1S and 1C_1C. For this one-layer solution, there are 737 second layer solutions for 1S (1S_1S through 1S_737S) and three second layer solutions for 1C.

There are also 2735 solutions which have the parallel proportion in the number of pieces. The first example is shown in Table A.32.

Appendix A: More Parallel Proportion Results

Table A.31 Sei solo movements 1:1 two-layer example

	Index	1S	1C	1S_1S	1S_1C	1C_1S	1C_1C
	Layer	1	1	2	2	2	2
Piece	Length	1200	1200	600	600	600	600
1001/1	22	22		22			
1001/2	94	94		94			
1001/3	20	20		20			
1001/4	136	136		136			
1002/1	24	24		24			
1002/2	24	24		24			
1002/3	80	80		80			
1002/4	80	80		80			
1002/5	32	32		32			
1002/6	32	32		32			
1002/7	68	68			68		
1002/8	68	68			68		
1003/1	23	23			23		
1003/2	289	289			289		
1003/3	26	26			26		
1003/4	58	58			58		
1004/1	32	32		32			
1004/2	54		54			54	
1004/3	29		29			29	
1004/4	40		40				40
1004/5	257		257			257	
1005/1	47	47			47		
1005/2	354		354				354
1005/3	21	21			21		
1005/4	102		102			102	
1006/1	138		138				138
1006/2	24	24		24			
1006/3	92		92			92	
1006/4	34		34			34	
1006/5	32		32			32	
1006/6	36		36				36
1006/7	32		32				32

Table A.32 Sei Soli example of double parallel 1:1 in two layers of bars and pieces

Index	32447S	32447C	32447S_1S	32447S_1C	32447C_1S	32447C_1C
Layer	1	1	2	2	2	2
Sum	1200	1200	600	600	600	600
Count	16	16	8	8	8	8
1001/1	22		22			
1001/2	94		94			
1001/3	20		20			
1001/4	136		136			
1002/1	24			24		
1002/2	24			24		
1002/3	80			80		
1002/4		80			80	
1002/5		32			32	
1002/6		32			32	
1002/7	68		68			
1002/8	68		68			
1003/1		23			23	
1003/2		289				289
1003/3	26			26		
1003/4		58				58
1004/1		32				32
1004/2	54		54			
1004/3		29				29
1004/4		40			40	
1004/5		257			257	
1005/1	47			47		
1005/2	354			354		
1005/3	21			21		
1005/4		102			102	
1006/1	138		138			
1006/2	24			24		
1006/3		92				92
1006/4		34			34	
1006/5		32				32
1006/6		36				36
1006/7		32				32

Appendix A: More Parallel Proportion Results

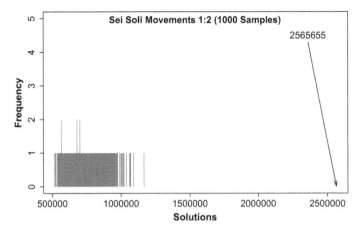

Fig. A.16 Monte Carlo simulation of Sei Soli movements 1:2

1:2 Proportion

Taking the 32 individual movements, there are 2,565,655 ways of obtaining a 1:2 proportion. Of these, 47 have a mirrored left/right symmetry and 41 have the same sequence in the left and right halves.

The solution given in Tatlow (2015) (p. 135) is the 2,565,583rd solution found by the program with 1:2 or the 73rd one found if it is run with 2:1.

A second layer is not possible.

The Monte Carlo simulation with 1000 samples again gives less solutions than the real collection, with an average of 732,423 and a maximum of 1,165,470 solutions—the distribution is shown in Fig. A.16.

Again, with the large spread in the numbers of solutions, the distribution is almost uniform with nearly all occurring only once.

There are no second layer solutions for 1:2 because 800 and 1600 are not divisible by 3.

Summary

Taking the works as a collection of six pieces shows no 1:1 and one 1:2 proportion. The Monte Carlo simulations show that 1:2 is about twice as likely as 1:1.

Three of the works show proportions between their movements.

Taking the movements as a set gives millions of solutions with several symmetricalpatterns. See Table A.33. There are many two-layer solutions for 1:1.

The Monte Carlo simulations could initially only be run with 1000 samples rather than the 100,000 we would have liked. These show that equivalent random sets also have millions of solutions, but not as many as the real collection. A later run with 5000 samples brings more solutions to light, but still not nearly as many as the real collection. The flat shape of the distribution can be attributed to the small number of samples being used.

Table A.33 Proportions in Sei Soli

	Works		Movements one layer			Movements two layers (Strict)		
	Solutions	Symmetries	Solutions	Left=Right + Left=thgiR	Proportion count	Solutions	Left=Right + Left=thgiR	Proportion count
1:1	0	0	2,602,210	39 + 39	445,852	2,501,937	0	2,735
1:2	1	0	2,565,655	41 + 47	0	–	–	–

A.17 Transcribed Concertos BWV 972–980, 592, 981–982

Bach transcribed orchestral concertos by Vivaldi and other Italiancomposersfor organand harpsichord during his time in Weimar 1703 and 1708–1717. He is thought to have learned important elements of his style from these.

The works have relatively few solutions and no symmetricalpatterns (Table A.34).

The Monte Carlo simulations (Fig. A.17) show that the real numbers of solutions are in much lower percentiles than other collections. If we assume firstly that Bach worked with proportions and secondly that the lengths tend to be chosen to give more solutions than a random collection, this could indicate that proportions were not intended here. This is plausible as Bach did not compose these works but copied and arranged them for organ in order to study them. He would have had no reason to adjust the lengths to obtain proportions.

Table A.34 Proportions in transcribed concertos

	Works			Works with repeats		
	solutions	Left=Right + Left=thgiR	Proportional count	Solutions	Left=Right + Left=thgiR	Proportional count
1:1	2	0	2	1	0	0
1:2	–	–	–	3	0	1

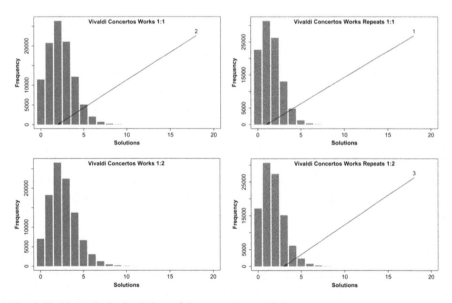

Fig. A.17 Monte Carlo simulations of the concerto transcriptions

Table A.35 Partial result for Vivaldi's L'Estro Armonico

Table	Pieces	Solutions	Combinations	Tried	% tried	Prob. Solution (%)
1:1	40	510,482,156	1,099,511,627,775	160,924,958,001	14.64	0.32
1:2	40	1,001,977,040	1,099,511,627,775	214,386,371,315	19.50	0.47

This does knot say anything about whether Vivaldi used proportions, as this is not a collection of Vivaldi's works exclusively. Bach's first biographer Forkel implies that Bach learned something about good proportions from Vivaldi.

If we look at the original Estro Armonico Op. 3, there are a few movements with repeats and four which have anacruses. The anacruses are not balanced by a shorter bar at the end, so one must decide whether to count them. (Bach apparently did not correct these in his transcriptions.) The lengths[2] give the following picture:

Counting repeats	Counting anacruses	Length	Possible proportions
N	N	2894	1:1
N	Y	2898	1:1, 1:2
Y	N	3093	1:2
Y	Y	3097	–

Exploring the 40 movements of L'Estro Armonico would take too long and so has not been attempted in full, but running the program for just over 24 h showed that this is no different to any of Bach's works. (This is counting without repeats and with the anacruses.) Table A.35 shows the percentage of the possible combinations tried and the resulting probability of finding a solution and these can be compared with Table 10.22.

A.18 Trio Sonatas for Organ BWV 525–530

These pieces, written in Leipzig in the late 1720s, were instruction pieces for Bach's son Wilhelm Friedemann. Although they only have one voice on each hand and the pedals, they are quite difficult to play as each voice is an independent part.

The Works

Taking the six works, there are no 1:1 or 1:2 proportions whether taken with or without repeats.

The Movements

The movements do give some proportions when counted with or without repeats.

---

[2] Using the Eulenburg Study Score edited by Christopher Hogwood, 2002.

Appendix A: More Parallel Proportion Results

Without repeats, we have 254 solutions for the 1:1 proportion of 780:780 bars, the first of which is the one given in Tatlow (2015) (p. 286). This also has a doubleproportion in the numbers of pieces 9:9, and there are 99 such double proportion solutions in all. There are two Left=Right symmetrical patterns shown in Table A.36 with the BWV numbers along the top and the lengths in bars divided between solution and complement in the other pairs of rows. Tatlow (2015) gives a second layer of proportion with 390:390 bars, and there are 31 of these. Tatlow (2015) also gives a third layer with 195:195 bars; we do not regard this as a strict solution because the other branch of layer 2 does not have any solutions. There are no strict three-layer solutions. In terms of the program output, Tatlow (2015) is giving solutions 1S and 1C, 1S_3S and 1S_3C, 1C_1S_1S and 1C_1S_1C.

Table A.37 shows the first two of the 254 solutions.

The overall results are summarised in Table A.38.

Counting with repeats the picture is similar—see Table A.39. Tatlow (2015) unusually gives a 2:3 proportion, but this splits the repeat of BWV 527/2 between the solution and complement, so is not regarded as a valid solution by the program.

The Monte Carlo simulations show that the real numbers of solutions are not as far above the average of the simulations as some other collections. Only the 1:1 and 1:2 without repeats are in the 93rd percentile and the 1:1 with repeats is in the 50th percentile, in other words about the same as the averageof random collections. These are shown in Fig. A.18—without repeats on the left and with repeats on the right.

Table A.36 Left=Right symmetries in Trio Sonatas without repeats

Index	Pattern	525/1	525/2	525/3	526/1	526/2	526/3	527/1	527/2	527/3	528/1	528/2	528/3	529/1	529/2	529/3	530/1	530/2	530/3
111S	Left=Right+^	58	28			48		112	32	144	64	45					180	40	77
224S	Left=Right+^	58					172			144	64				54	163			77

Appendix A: More Parallel Proportion Results

Table A.37 First solutions of Trio Sonatas without repeats

Index	Layer	525/1	525/2	525/3	526/1	526/2	526/3	527/1	527/2	527/3	528/1	528/2	528/3	529/1	529/2	529/3	530/1	530/2	530/3	Sum	Count
1S	1	58	28	64	78	48	172	112									180	40		780	9
1C	1								32	144	64	45	97	104	54	163			77	780	9
1S_1S	2	58		64		48											180	40		390	5
1S_1C	2		28		78		172	112												390	4
1S_2S	2	58				48	172	112												390	4
1S_2C	2		28	64	78												180	40		390	5
1S_3S	2	58			78		172	112												390	4
1S_3C	2		28	64	78	48	172													390	5
1C_1S	2								32		64				54	163			77	390	5
1C_1C	2									144		45	97	104				40		390	4
1C_1S_1S	3								32						54	163				195	2
1C_1S_1C	3										64	45	97		54					195	3
2S	1	58	28	64	78	48	172	112	32	144	64	45				163	180	40	77	780	12
2C	1							112		144	64	45	97	104	54				77	780	6
2S_1S	2	58	28		78		172		32				97		54					390	5
2S_1C	2			64		48					64	45	97					40		390	7
2S_2S	2	58		64	78	48			32		64	45								390	6
2S_2C	2		28				172		32		64				54			40		390	6
2S_3S	2	58		64	78		172		32			45			54			40		390	7
2S_3C	2		28			48	172				64		97							390	5
2S_4S	2	58		64			172		32		64									390	5

(continued)

Table A.37 (continued)

Index	Layer	525/1	525/2	525/3	526/1	526/2	526/3	527/1	527/2	527/3	528/1	528/2	528/3	529/1	529/2	529/3	530/1	530/2	530/3	Sum	Count
2S_4C	2		28		78	48						45	97		54			40		390	7
2S_5S	2	58		64					32			45	97		54			40		390	7
2S_5C	2		28		78	48	172				64									390	5
2S_5S_1S	3	58							32			45	97		54			40		195	3
2S_5S_1C	3			64								45	97							195	4
2S_6S	2	58			78	48					64	45			54					390	6
2S_6C	2		28	64			172		32						54			40		390	6
2S_7S	2	58							32		64	45	97		54			40		390	7
2S_7C	2		28	64	78	48	172													390	5
2S_7S_1S	3	58											97					40		195	3
2S_7S_1C	3								32		64	45			54					195	4

Appendix A: More Parallel Proportion Results

Table A.38 Proportions in Trio Sonatas movements without repeats

	Movements one layer			Movements two layers (Strict)		
	Solutions	Left=Right + Left=thgiR	Proportion count	Solutions	Left=Right + Left=thgiR	Proportion count
1:1	254	2 + 0	99	31	0	0
1:2	237	0 + 2	72	–	–	–
2:3	364	0	–	–	–	–

Table A.39 Proportions in Trio Sonatas movements with repeats

	Movements one layer			Movements two layers (Strict)		
	Solutions	Left=Right + Left=thgiR	Proportion count	Solutions	Left=Right + Left=thgiR	Proportion count
1:1	204	0 + 1	75	51	0	0
1:2	203	0	72	–	–	–
2:3	323	0 + 1	–	–	–	–

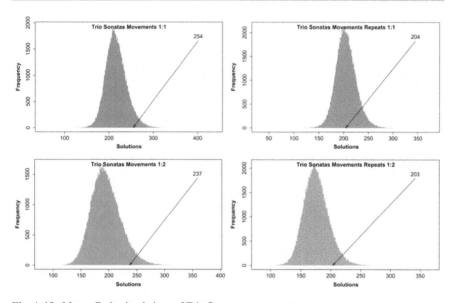

Fig. A.18 Monte Carlo simulations of Trio Sonatas movements

Table A.40 Proportions in Violin Sonatas works

	Works		Works with repeats	
	Solutions	Left=Right + Left=thgiR	Solutions	Left=Right + Left=thgiR
1:1	0	0	0	0
1:2	–	–	1	0

A.19 Violin Sonatas BWV 1014–1019

These sonatas for violin and harpsichord were written in Köthen up to 1723 and continually modified in the Leipzig years.

The Works

For the six works without repeats, there are no solutions for 1:1 and 1:2 is not possible.

For the works with repeats (Table 5.8 in Tatlow (2015)), there is no 1:1 proportion and only the one 1:2 proportion, which also has the double 1:2 proportion in the number of pieces (sonatas 3 and 5 against 1, 2, 4 and 6).

The Movements

Taking all the individual movements as a set (with and without the repeats) gives significantly more possibilities. The simpler proportions are summarised in Tables A.41 and A.42. Without repeats, there can only be a single layer of 1:1 and no 1:2 proportions. With repeats, there could be 1:1 with multiple layers and 1:2

Table A.41 Proportions in Violin Sonatas movements

	Movements without any error messages repeats			Movements without repeats two layers (Strict)		
	Solutions	Left=Right + Left=thgiR	Proportion count	Solutions	Left=Right + Left=thgiR	Proportion count
1:1	32,304	6 + 9	–	–	–	–
1:2	–	–	–	–	–	–
2:3	44,976	10 + 4	13,048	–	–	–

Table A.42 Proportions in Violin Sonatas movements with repeats

	Movements with repeats single layer			Movements with repeats two layers (Strict)		
	Solutions	Left=Right + Left=thgiR	Proportion count	Solutions	Left=Right + Left=thgiR	Proportion count
1:1	23,003	4 + 7	–	19,288	0	–
1:2	18,182	5 + 5	–	–	–	–
2:3	33,107	6 + 10	8,428	29,339	402 + 43	4,446

Appendix A: More Parallel Proportion Results

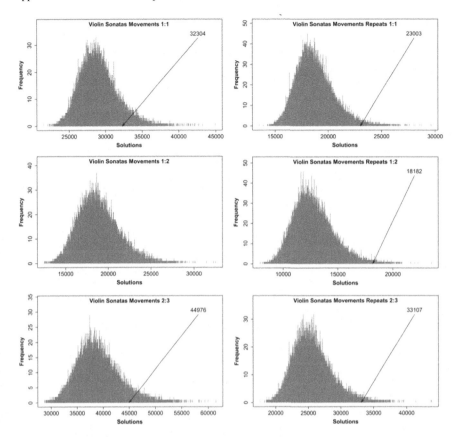

Fig. A.19 Monte Carlo simulations of Violin Sonatas movements 1:1

with one layer. We also include the 2:3 proportions as these produce many with the doubleproportion in the number of pieces.

The works or movements with repeats would also give other proportions such as 1:3, 1:4, 1:5, 1:7, etc., as the total length of 2400 has many factors.

The histograms in Fig. A.19 show the distributions of the Monte Carlo simulations without repeats (left-hand side) and with repeats (right-hand side). The actual lengths for the 1:1 proportion are in the 93rd and 99th percentile, respectively. For the 1:2 with repeats, the actual lengths fall in the 99th. For 2:3, they are in the 95th without repeats and the 99th with repeats.

A.20 Well Tempered Clavier Book 1—See Chap. 9

This was covered in Chap. 9.

Appendix B
Proportional Parallelism Explorer Program User Manual

B.1 Installing the Program

B.1.1 Download

The program itself can be downloaded from:

https://github.com/Goldberg53/PPExp with no guarantees that it will work as expected.

Instructions are on that page.

If you are reading this in the book: the manual included with the program may contain updates made after the book was published.

If you are reading this from the program: this manual is an extract from the book "Let's Calculate Bach" and may have been updated since the book was published. Familiarity with the book is required in order to fully understand the program.

B.1.2 Prerequisite

The program is written in the Java programming language. The Java runtime environment (JRE) version 8 (also known as 1.8) or later must be installed before the program can run.

Obtain the official free JRE or JDK from Oracle at www.java.com and install it as instructed.

B.1.3 Caveats

The program has been tested on the following systems:

V6.0 on Apple MacOs Catalina 10.15.6
V5.0 on Microsoft Windows 10
V5.0 on Apple MacOS Mojave 10.14.3
V5.0 on Linux Ubuntu Xenial 16.04, Kernel 4.15.0–45-Generic
V4.0 on Microsoft Windows 10, Windows 7
V4.0 on Apple MacOS High Sierra Version 10.13.6
V3 on on Linux Ubuntu Xenial 16.04, Kernel 4.10.0–42-Generic.

The program has been tested, but as with any software, there can be no guarantee that all results are correct, and no responsibility is taken for any harm caused by using the program.

The following differences between the systems are known:

- Apple MacOS has tighter security restrictions on downloading and running unsigned files (see Sect. B.2). The program file is not signed, so the user must confirm download and execution.
- Apple MacOS does not implement the colouring of the progress bars for Pause and Cancel.

B.1.4 Installing the Program

The program does not need to be installed as an application. Download the.jar file to a convenient place on your computer. You can create a shortcut to the jar file and copy the shortcut to the desktop.

The user manual is included in the.jar file and can be extracted and opened with the "Extract Manual" function in the Help menu. Further information is in the Help function in the Help menu.

B.1.5 Removing the Program

To remove the program from your computer, simply delete the .jar file.

If you extracted the user manual, you can delete the extracted file in your home folder, called PropParExpManualV5.0.pdf or similar at any time.

If desired, uninstall the Java runtime environment in the usual way, e.g. on Windows with Programs and Features in the Control Panel.

B.2 Running the Program

B.2.1 From the Downloaded File

To run the program:

Windows: double click on the **.jar** file or the shortcut (see Sect. B.1.4).

It can also be started from a command line—see Sect. B.2.2.

Mac OS: double click on the **.jar** file or the shortcut (see Sect. B.1.4).

The program is not signed, so the Mac system may not allow it to run straight away, giving the message below:

This can be circumvented by control-clicking on the jar file and selecting "Open", which will then ask if you want to open it.

Alternatively, you can go to System Preferences—Security and Privacy and in "Allow apps downloaded from", select "Open Anyway" as shown below:

There may also be issues with access to the disc. It may be necessary to go to System Preferences—Security and Privacy and in Full Disc Access add the Java program, and for starting from the terminal add the Terminal program.

Unix: run from the command line (see Sect. B.2.2).

Multiple instances of the program can be run simultaneously.

B.2.2 Running the Program from the Command Prompt

To run the program from a command prompt:

Start a command prompt, e.g. in the Windows start menu under Windows System, on Mac OS start a terminal window.
Change the directory to the folder[3] where the PropParExp.jar file is saved.
Type the following command:
`java -jar PropParExpV6_0.jar` (or the file name with the current version).

The program should run and give any error messages in the command line window.

---

[3] The terms "directory" and "folder" are synonymous—the usage depends on the operating system.

The program does not support reading arguments from the command line, so it cannot be used for batch processing. (However multiple instances can be run simultaneously.)

B.2.3 Running Multiple Instances

The program can be run more than once to perform analysis of multiple file simultaneously or to run different Mont Carlo simulations at the same time. This is particularly useful if multiple runs are needed which take a long time and are to be run over night or even over several days.

If you are lucky enough to have a computer with multiple processor cores, each instance will run in a separate core, and will therefore genuinely run simultaneously. If you run more instances than you have cores, they will compete for the use of the available processors and take longer.

Note that running a file analysis and a Monte Carlo simulation in the same program instance will run these simultaneously on two processors if available. It is not necessary to start two instances of the program to do this.

Note also, that if you wish to do anything else while the program is running, such as reading mails or editing documents, this will be significantly slower unless you leave one processor core free for this work.

Windows 10—simply double click on the.jar file to start a new instance or from a command prompt type.

```
start java -jar PropParExpVm_n.jar
```

(this will also start a new terminal window—if you close this it will terminate the program as well).

MacOS—from a terminal command line window

```
open -n PropParExpVm_n.jar
```

or open a new Terminal window to start each instance with the command

```
java -jar PropParExpV6_0.jar
```

or start them asynchronously from a single terminal window by adding an ampersand to the command:

```
java -jar PropParExpV6_0.jar &
```

The most convenient method is to create a script file—see Sect. B.2.4.

Note that using the `open -n` command has the locale problem described in Sect. B.2.6.

B.2.4 Starter Script

To avoid having to open a command line or terminal to start the program in the above circumstances, you can make a script which will run with a double click.

Windows

```
Start java -Duser.language=de -jar….
```

There is no simple way to avoid leaving the command line window open.

MacOS

Create a file named e.g. RunPPE.command with the line
```
java -jar /Users/<you>n/PropParExpV6.0.jar &
```

Make the file executable with
```
chmod +x /Users/<you>/RunPPE.command
```

This will leave the terminal window open. To automatically close the terminal window when the program exits, go to Terminal—Preferences—Profiles—Shell and set "When the shell exits" to "Close if the shell exited cleanly".

B.2.5 CSV Separator

When it is started, the program pre-sets the CSV-file separator depending on the decimal separator:

- If the decimal separator is set to comma, the CSV separator is set to semicolon.
- If the decimal separator is dot or anything else, the CSV separator is set to comma.

This is the same logic as is used by Excel, so the CSV files should open in Excel with no problem (unless the system settings have been changed since the file was produced). Note also the problems with mixed languages—see Sect. B.2.6.

The separator can be changed in the preferences—see Sect. B.8.

B.2.6 Mixed Languages

If you have your computer set to a different language than your region, e.g. the operating system is in English but you are in Germany, the decimalnumber formatting may not work as expected. Java seems to define its decimal separator from the system language rather than the region locale setting, so even though your regional settings show comma as the decimal separator, it may use dot. The workarounds for this, apart from changing the system language, are as follows:

Windows 10

Start from the command line and use

    java -Duser.language=de -jar….

MacOS

Start from the terminal with

    java -jar…

and if you need multiple instances, start a new terminal window for each, as open will not use the correct setting.

CSV Separator

This also depends on the language settings—see Sect. B.2.5.

Thousands

The thousands separator has been set to space to aid legibility and avoid any confusion with decimal and CSV separators.

B.2.7 Error Messages

Errors handled by the program will give a message on the screen, e.g. if the output file could not be created.

If you need assistance in solving the problem, e.g. if the message contains a cryptic stack dump, send a screenshot of the message to the developer.

If the program fails without an error message, or more information is required for the developer, start it from a command prompt (see Sect. B.2.2) in order to see the console output, and repeat the action. This is useful as internal error messages are displayed in the console, which is not otherwise visible.

Various error messages may arise from checking the input file. These include the line number of the input file so that they can easily be localised.

An error "java.lang.UnsupportedClassVersionError" means that you have an older version of Java (see Sect. B.1.2).

B.3 File Processing

B.3.1 File Processing Data Flow

The basic data flow for processing an input file to find the proportions is:

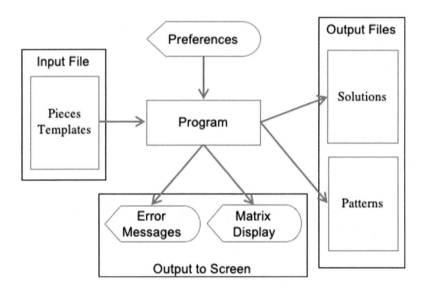

The program works as follows:
- In the preferences dialog, the user sets the desired proportion, the contents of the output files and other parameters (see Sect. B.8).
- When the user opens an input file, read the file to get the titles and lengths (number of bars) of a set of pieces and the pattern matching criteria (templates)—see Sects. B.3.2–B.3.7; check the input for errors and output them to the screen—see Sect. B.2.7.
- When the user starts the file processing, search for combinations of the pieces that will satisfy a given proportional split, i.e. Solutions; errors occurring during processing are output to the screen (see Sect. B.2.7).
- For each solution found:
    - The solutions are optionally shown on the screen in real time in a matrix display (see Sect. B.8).
    - Write the solution and optionally its complement to the solutions output file (see Sect. B.3.8).
    - Test each solution pair against all the patterns given in the input file.

- Write solutions with matching patterns to the patterns output file (see Sect. B.3.9).
- Write the summary to the output files and the screen.

B.3.2 Input File—General

The input file consists of different types of line, each with their own syntax, for defining a set of pieces and patterns to match.

Each line starts with a keyword denoting the type of line. The keywords are not case sensitive, i.e. capital letters can be used at will.

The elements of a line can be separated by any number of spaces or tabs.

Empty lines are ignored.

Comment lines starting with // are ignored and can be used for your own notes.

A line can conclude with a comment starting with //.

It is recommended to keep the name of the input file short, as various items are appended to it to form the output file names—see Sect. B.8.

Input files can be created with a text editor (e.g. Notepad on Windows or TextEdit on Mac) or with a text processing program such as Microsoft Word by saving the file as plain text.

B.3.3 Input File—Pieces

The set of pieces must be provided in the input file with one piece per line.

The syntax of the lines is as follows (angled brackets denote a placeholder for variable content):

Piece <title> <length>

Each line consists of the keyword "Piece", the title of the piece and its length in bars as an integer number.

The title cannot contain spaces, as these delimit the elements of the line—an underline or hyphen can be used instead. The title is used for column headers in the output files.

There cannot be more than 63 pieces in a file (this is limited by the size of the largest integer 2^{63} when counting the combinations).

Example for J. S. Bach's Well Tempered Clavier Book 1:

```
// J. S. Bach Well Tempered Clavier Book 1
Piece     01_Cmaj_a4               62
Piece     02_cmin_a3               69
Piece     03_C#maj_a3              159
Piece     04_c#min_a5              154
Piece     05_Dmaj_a4               62
Piece     06_dmin_a3               70
Piece     07_Ebmaj_a3              107
Piece     08_d#min_a3              127
Piece     09_Emaj_a3               53
Piece     10_emin_a2               83
Piece     11_Fmaj_a3               90
Piece     12_fmin_a4               80
Piece     13_F#maj_a3              65
Piece     14_f#min_a4              64
Piece     15_Gmaj_a3               105
Piece     16_gmin_a4               53
Piece     17_Abmaj_a4              79
Piece     18_g#min_a4              70
Piece     19_Amaj_a3               78
Piece     20_amin_a4               115
Piece     21_Bb(B)maj_a3           68
Piece     22_bb(b)min_a5           99
Piece     23_B(H)maj_a4            53
Piece     24_b(h)min_a4            123
```

B.3.4 Input File—Patterns: Overview

The pattern matching is described in more detail in the book Let's Calculate Bach.

There are three functions, Template, Count and BuiltIn, described in the following sections. The lines can be in any order. Each solution pair found, i.e. each solution and its complement, is tested against all the given patterns.

B.3.5 Input File—Patterns: BinaryTemplate

Binarytemplates can be given to either exactly match the signature or match a part of the signature in any position.

These are defined by lines in the input file with the following syntax:
    Template <name> <action> <binarytemplate>

Appendix B: Proportional Parallelism Explorer Program User Manual   293

The keyword "Template" denotes the type of line (as opposed to "Piece"—see Sect. B.3.3).

\<name\> is a name for this template which the user can define. This is shown in the output file (see Sect. B.3.9). It cannot contain spaces, as these delimit the elements of the line—an underline or hyphen can be used instead.

\<action\> defines how the pattern is to be matched. This can be "Exact" or "Shift".

\<binarytemplate\> is a sequence of bits to be matched against the solution signatures.

If the template has the wrong number of bits, an error message is shown. Comment lines showing the bit positions are useful for checking that one has the correct number of bits and they are lined up as desired.

The following is a simple example of templates in an input file.

```
// Pattern Templates
// Bit Positions 10s        000000000111111111122222
// Bit Positions 1s         123456789012345678901234
Template    Test      Shift 1111111001
Template    Tatlow    Exact 111110000001000000011111
Template    12inARow  Shift 111111111111
```

B.3.6 Input File—Patterns: Count

This pattern will match a solution that consists of a given number of pieces. It is defined by lines in the input file with the syntax:

Count \<name\> \<number\>

The keyword "Count" denotes the type of line (as opposed to Piece or Template).

\<name\> is a name for the pattern as for binarytemplates.

\<number\> is an integer giving the number (or count) of pieces in a solution or its complement which will give a match.

The following example will give a match if a solution or its complement is made up of any 12 pieces or 8 pieces, respectively.

```
Count Twelve      12    // Twelve anywhere
Count Eight       8     // Eight anywhere
```

Note that all solutions are checked for having the proportion in the number of pieces—see PropCount in Sect. B.3.8.

B.3.7 Input File—Patterns: Built in Functions

The "built in" patternmatching functions are denoted as follows by lines in the input file with the syntax:

        BuiltIn   <function>

The functions are predefined, and the following are currently available.

Left=Right
        BuiltIn   Left=Right

- this finds any solutions where the right half of the signature is the same as the left half, e.g.

        010100 010100

Left=thgiR
        BuiltIn   Left=thgiR

- this finds any solutions where the right half of the signature is the mirror image or reverse of the left half, e.g.

    010100 001010

The following example shows both of these added to the previous example:

```
Count Twelve      12         // Twelve anywhere
Count Eight        8         // Eight anywhere
// Pattern Templates
// Bit Positions 10s                    00000000011111111122222
// Bit Positions 1s                     123456789012345678901234
Template    Test           Shift 1111111001
Template    Tatlow         Exact 111110000001000000011111
Template    12inARow       Shift 111111111111
BuiltIn     Left=Right
BuiltIn     Left=thgiR
```

If there is an odd number of pieces, the middle piece is irrelevant for the above, e.g. Left=Right will match both
0101000 010100 and
010100 1 010100

B.3.8 Solutions Output File

The solutions output is delivered to a file either in the same folder as the input file or a different folder chosen by the user (for example if all output files are to be gathered in the same folder). The details are determined by the preferences—see Sect. B.8.

The first line consists of the titles of the columns, with the titles of the pieces as given in the input file from column F (this can be omitted—see Sect. B.8).

The basic output is a sequence of pairs of lines each of which lists a combination of lengths which will result in the given proportion, i.e. the solution pairs. The columns are as follows:

Index—the sequential index number of the solution pair. The two solutions of a pair have the same index, suffixed with an S for solution and C for complement, and this is referenced in the patterns output file (Sect. B.3.9). Each layer has an additional layer in the sequence number, separated by underlines. The complement lines are omitted if the preference "Include Complements" (see Sect. B.8) is not checked—this is the initial setting.

Layer—this gives the layer number of the solution pair, which is also indicated in the Index.

Sum—this column gives the sums of the lengths of the solutions and complement.

Count—this column gives the number of pieces in each solution or complement row.

PropCount—this column indicates whether the number of pieces is in the same proportion as that chosen for the lengths, i.e. that it is a double proportion.

Note that for the 1:1 proportion there is a specific setting to determine whether the opposites are included.

Note that the sum and count columns can be excluded in the preferences.

	A	B	C	D	E	F	G	H	I	J	K
1	Index	Layer	Sum	Count	PropCount	U	V	W	X	Y	Z
2	1S	1	30	3	Y	5	10	15			
3	1C	1	30	3	Y				8	7	15
4	1S_1S	2	15	2		5	10				
5	1S_1C	2	15	1				15			
6	1C_1S	2	15	2					8	7	
7	1C_1C	2	15	1							15
8	2S	1	30	4		5	10		8	7	
9	2C	1	30	2				15			15
10	2S_1S	2	15	2	Y	5	10				
11	2S_1C	2	15	2	Y				8	7	
12	2C_1S	2	15	1	Y			15			
13	2C_1C	2	15	1	Y						15
14	3S	1	30	3	Y	5	10				15
15	3C	1	30	3	Y			15	8	7	
16	3S_1S	2	15	2		5	10				
17	3S_1C	2	15	1							15
18	3C_1S	2	15	1				15			
19	3C_1C	2	15	2					8	7	
20	Program Version: 6.0d										
21	Input File: /Users/alan/OneDrive/PropParResults/Z_Book/6test_for book.txt.										
22	Total Length: 60. Proportion: 1:1. Target: 30. Tried: 38.										
23	Include Complements: Yes. 1:1 Opposites: No.										
24	Layers: 2. Strict Layers: No. Strict PropCount: No. Only First Strict: No.										
25	Strict Proportion Solutions: 3.										
26	Strict PropCount Solutions: 0.										
27	Solutions Layer 1: 3. Layer 2: 6.										
28	Patterns: 1.										
29	Processing Time: 0,001 Seconds without Matrix.										

The file concludes with a summary of the results—see Sect. B.3.10 for details.

B.3.9 Patterns Output File

A second output file is created which gives those solutions which match one or more of the patterns defined in the input file (see Sects. B.3.4–B.3.7). This is written to the same folder as the solutions file.

	A	B	C	D	E	F	G	H	I	J
1	Index	Layer	Pattern	PropCount	U	V	W	X	Y	Z
2	2S	1	Left=Right+^		5	10		8	7	
3										

These are shown as follows:

Index—a reference to the index number of the solution pair in the solutions output file (see Sect. B.3.8).

Table B.1 Pattern output indicators

Indicator	Meaning
>n	For shift template: the pattern was shifted right by n places to match
ˆ	The pattern matches the complement of the given solution Notes This can occur in addition to the above as >nˆ, i.e. the pattern shifted n places matches the complement The pattern output gives the lengths for the solution being tested (the first row of the pair in the solutions file) and not its complement where the match occurred
+ˆ	The pattern matches both the signature and its complement For the built in patterns Left=Right and Left=thgiR the pattern that matches the signature will always match the complement as well, so this is always present

Layer—the layer in which the pattern occurs.

Pattern—the name of the pattern as given in the input file. This is suffixed with an indicator for where the pattern occurred as shown in the Table B.1.

PropCount—shows "Y" for yes if there is an additional proportion in the count of pieces.

A solution pair can match multiple patterns. This will be apparent as the same index will occur in consecutive rows with different patterns in the pattern output file.

B.3.10 File Processing Summary

The summary contains:

- The program version with which the results were obtained.
- Whether the run was cancelled before completion
- The input file name with path.
- The total length of all the pieces in bars, the chosen proportion, the target length for that proportion and the number of combinations actually tried.[4]
- The preference settings of Include Complements and 1:1 Opposites.
- The number of layers selected and, if more than one, the settings for Strict Proportions and Only First and for Proportional Counts and StrictProportional Count.
- The number of strict solutions found (if more than one layer).
- The number of strictproportional count solutions found (if more than one layer).
- The number of solutions and the number of PropCounts found in each layer.
- The number of patterns found.

---

[4]Note: this will usually be less than the theoretical number of possible combinations, as the program is optimised to stop trying a sequence when the target is exceeded—see the book for details.

- The time taken, indicating with or without matrix output, as this slows the program down significantly.

Note: The Proportional Counts show "-" if the number of pieces in the layer cannot be divided in the desired proportion.

The summary is shown in a window on the screen when the program completes, at the end of the matrix output and optionally at the end of the solutions output file.

B.4 Monte Carlo Simulation

B.4.1 Monte Carlo Processing Data Flow

For Monte Carlo simulation the process is shown below and is as follows:

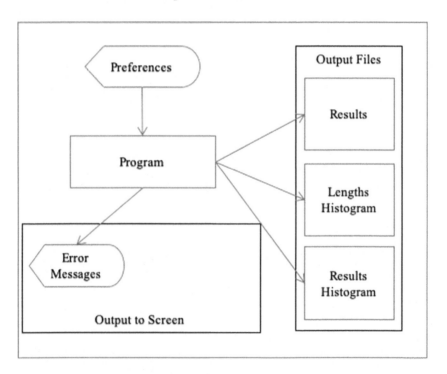

- The user sets the desired proportion, the number of pieces, number of samples, minimum and maximum length and the folder for the output file and other parameters in the preferences dialog (see Sect. B.8).
- When the user starts the simulation, generate pseudo-random sets of lengths.
- For each sampleset:
  - Run the search routine to give the number of solutions

- Write the lengths and number of solutions to the results output file.
- Write the summary to the output file and optionally output the histogram files.

The histogram files are intended to facilitate plotting graphs of the results, e.g. with Excel or the Rstatistics package.

B.4.2 Results Output File

The output file name is determined by the preferences—see Sect. B.8.

The output file has each sampleset of pieces in a separate row. The columns are simply titled "R1", "R2", etc. for "Random 1" etc. The next columns are the number of solutions for each layer. The last column is the time taken to find the number of solutions.

The file includes a summary—see Sect. B.4.3 for details.

Note: the actual solutions for each sample are not output. If these are required, the user must take the lengths of a sample and create an input file as described above.

B.4.3 Monte Carlo Summary

The summary is written at the end of the samples output file and shown in a window on the screen when the program completes. It contains:

- The program version with which the results were obtained.
- Whether the run was cancelled before completion.
- The number of samples tried and the number of samples ignored (because they cannot give the desired proportion).
- The number of pieces and number of samples taken.
- The minimum and maximum lengths of the pieces.
- For each layer:
  - The minimum and maximum numbers of solutions.
  - The mean and standard deviation of the numbers of solutions.
- The time taken.

	A	B	C	D	E	F	G	H	I	J	K
1	R1	R2	R3	R4	R5	R6	L1_Bars	L1_Pieces	L2_Bars	L2_Pieces	Seconds
2	15	14	10	8	6	3	1	1	1	0	0
3	14	11	8	4	4	3	2	1	2	0	0
4	13	11	7	3	3	3	1	0	0	0	0
5	15	13	12	8	4	4	1	0	0	0	0
6	15	14	11	10	9	5	0	0	0	0	0
7	10	7	6	5	4	4	1	1	0	0	0
8	15	15	14	13	10	9	2	2	0	0	0
9	14	12	12	9	6	3	0	0	0	0	0
10	14	10	9	7	5	3	2	1	1	1	0
11	14	13	13	10	3	3	0	0	0	0	0
12	15	12	10	9	9	5	1	1	1	0	0
13	14	14	11	0	0	2	0	0	0	0	0
99998	13	13	10	8	5	3	3	2	6	2	0
99999	14	11	11	7	5	4	1	1	0	0	0
100000	11	8	6	6	5	4	1	1	0	0	0
100001	15	14	13	12	10	8	1	1	0	0	0
100002	Program Version: 6.0e
100003	Tried: 100 000. Ignored: 300 554.
100004	Pieces: 6. Samples: 100 000.
100005	Minimum Length: 3. Maximum Length: 15.
100006	Proportion: 1:1. Strict Layers: No. Strict PropCount: No. Layers: 2.
100007	Layer 1:
100008	Min. Solutions: 0. Max. Solutions: 7.
100009	Mean: 1,05005. Std. Deviation: 0,96921.
100010	Min. PropCounts: 0. Max. PropCounts: 6.
100011	Mean PropCounts: 0,84725. Std. Deviation: 0,91473.
100012	Layer 2:
100013	Min. Solutions: 0. Max. Solutions: 16.
100014	Mean: 0,40753. Std. Deviation: 0,98431.
100015	Min. PropCounts: 0. Max. PropCounts: 15.
100016	Mean PropCounts: 0,11787. Std. Deviation: 0,43708.
100017	Processing Time: 1,462 Seconds.

B.4.4 Lengths Histogram Output File

The output file name is determined by the preferences—see Sect. B.8.

The output file provides the data prepared for plotting a histogram of the lengths of all the pieces in all the samples. It consists of two columns, one for all the lengths in the range, and for each length, the number of pieces that have that length, i.e. the frequencies.

This can be used to validate that the pseudo-random numbers generated have an approximately uniform distribution.

B.4.5 Results Histogram Output File

The output file name is determined by the preferences—see Sect. B.8.

The output file provides the data prepared for plotting a histogram of the results. It consists of one column for the number of solutions and one for each layer with the number of samples that had that number of solutions, i.e. the frequencies.

Note: with Strict Proportions only one column is output for layer 1.

B.5 Further Processing with Other Programs

B.5.1 Import to Excel

The output files can be used in Microsoft Excel for further exploration, e.g. by sorting or filtering, or making histogram graphs.

To open the output file directly in Excel, it must have the correct separator, which can be determined as follows. See also Sect. B.2.6.

For Windows: in the Windows Control Panel, under Region, Formats tab and Additional Settings, the List Separator is shown. This depends on the operating system language setting and is usually comma in English[5] or semicolon in German.

On MacOS the CSV separator is comma if the decimal separator is dot and is semicolon if the decimal separator is comma.

The program tries to select the appropriate separator as the default.

Use the same separator in the program preferences and set the output file extension to csv—see Sect. B.8.

If the separator is set to Tab, you can open the file in a text editor, copy the content (ctrl-A, ctrl-C) and paste it into Excel.

For use with Excel you may wish to leave the "Include Summary" box unchecked, as the final summary lines do not fit the column structure and may be disturbing if sorting and filtering are used. However, it is recommended to leave it in if many files are being processed so that you can later verify the parameters that produced the file.

B.5.2 Formatting Histograms with Excel

To facilitate further analysis the data for histograms can be written to files (see Sects. B.4.4 and B.4.5).

For a simple plot in Excel, select the data and Insert a Column Chart. Format the data series, e.g. with no Series Overlap and a small Gap Width.

---

[5]Hence the name "Comma Separated Variables" for the csv file extension.

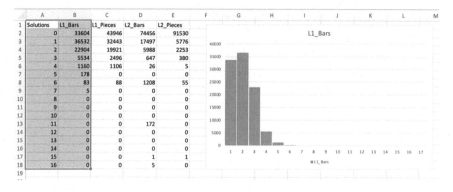

Note: the outputs of the histogram data are prepared for a direct x–y bar plot with x as the number of solutions and y number of samples that have x solutions (the frequency), and so are not suitable for use with histogramfunctions. We find this better for large numbers of samples because the standard histogramfunction combines the data into "bins", e.g. with 1–10 in the first column, 11–20 in the second column, etc. and the exact appearance of the diagram then depends on the bin size. (Excel is also limited to 1000 bins.) Using the data as provided by the program effectively has a bin size of 1 and shows the data accurately. To use an Excel histogramfunction on the results, copy the "Solutions" column of the main output file.

B.5.3 Text Processors

If the separator is set to Tab, you can copy the output from the file and paste it into a Word table.

B.6 Main Window

The main window appears as shown.

Appendix B: Proportional Parallelism Explorer Program User Manual

The program has four menus, Program, File, Monte Carlo and Help, with menu items described below.

Beneath this it shows for input file processing, the file that is open (if any) and the progress bar, and for Monte Carlo processing the progress bar.

The progress bars show a count of the number of solutions or samples tried and the maximum number, rather than a percentage. This is better for long running explorations which, taking many hours or even days to complete, remain at 0% for a long time. If the run is cancelled, the remaining part of the progress bar is red. If the run is paused, the remaining part is yellow. The progress bar titles also show the proportion when a process is started.

The file processing and Monte Carlo simulation can both be run at the same time.

The program can also be started multiple times, e.g. to leave several explorations running over night.

B.7 Program Menu

The program menu has:

- Set the preferences
- Close the program.

B.8 Preferences

The program is controlled by the preferences. This is divided into three tabs, General, Monte Carlo and Output.

Note that the close button (x) at the top right (or top left on Mac) is disabled. The preferences can only be closed with the OK button, to ensure that the error checking is performed.

If an error is found when the OK button is clicked, a message will appear on the screen and the tab containing the error will be shown. The preferences window will then not be closed, so that the error can be seen and corrected if necessary. If there are multiple errors, they will appear sequentially.

All the preferences are set to default values when the program is started.

The preferences can be opened at any time, even while the program is running. However, any changes to the parameters will only take effect on the next run.

B.5.4 General Tab

Shown here on Windows 10.

Appendix B: Proportional Parallelism Explorer Program User Manual

Proportion and Matrix

The desired **proportion**, e.g. 1:1, 1:2 is set with the two numbers. They are currently limited to a maximum of 99. They can be typed in or increased/decreased with the arrows beside the numbers.

If you choose a proportion which has common factors, it will be divided by the highest common factor, e.g. choosing 4:6 will use 2:3. This is indicated with a message on the screen.

If you choose a proportion which is not possible with the given set, e.g. 1:1 when the total number of bars is an odd number, a message will be displayed on the screen when the processing is started and no output file will be produced.

For proportions other than 1:1, putting the smaller number first (e.g. 1:2 rather than 2:1) will take less time to run, as it uses a smaller target value.

To obtain the oppositesolution pairs (other than for 1:1), use the reverse proportion, e.g. 2:1 instead of 1:2. To obtain the opposites of a 1:1 proportion, check the checkbox "**1:1 Opposites**"—the program will output all the solution pairs including the opposites. This checkbox is only visible when a 1:1 proportion is selected, as it is otherwise irrelevant.

Show Matrix

If this box is checked the results of file processing are scrolled into a new window on the screen.[6] This is useful if you wish to observe the search in real time. You may need to make the window wider to see all the columns.

Note that the program will take significantly longer if this option is selected, and the time taken given in the summary shows whether the time is with or without the matrix.

Layers

For the layers, the number of layers can be set between 1 and 9.

The Strict settings are only available for more two or more layers.

Strict Proportions will only find solutions that are complete over all the given layers.

Strict Proportional Counts is only available if Strict Proportions is checked and will find any solutions where the number of pieces in the solution (or its complement) is equal to:

$$\text{No. of pieces in solution} = \text{No. of pieces in collection} * \left(\frac{m}{m+n}\right)$$

with a proportion of m:n.

It will match solutions where the number of pieces is in proportion, e.g. for 24 pieces

---

[6]The output is intentionally reminiscent of the science fiction film "The Matrix", Warner Bros.

and 1:1—solutions with 12:12 pieces will match
and 1:2—solutions with 8:16 pieces will match
and 1:3—solutions with 6:18 pieces will match.

These are equivalent to using

```
Count Any12 12
Count Any8 8
Count Any6 6
```

respectively, but saves having to edit the input file when changing the proportion.

The pattern will also match the complement of a solution, which is the same as matching the "n" part of the m:n proportion with the solution.

Only First is also only relevant if Strict Proportions is checked and with this the program will only output the first solution found in each sublayer. This saves time and saves space in the output files.

B.5.5 Monte Carlo Tab

This tab is used to set the parameters for the Monte Carlo simulation. The numbers can be typed in or increased/decreased with the arrow buttons.

The entries are limited to:

Number of pieces: 2–63
Number of samples: 1–2,147,483,647 (a 32-bitinteger $2^{31} - 1$)
Minimum length: 1–1 000
Maximum length: 1–10 000.

The set button can be used to set the # Pieces, Minimum length and Maximum length and the prefix for output file names to the values from the input file that is currently open. The button is disabled (greyed out) until a file has been opened. (See Sect. B.8 for details of file naming.)

If the maximum length is less than the minimum length, when OK is clicked, the values will be swapped, the preferences window will be switched to the Monte Carlo tab and a message "Monte Carlo Lengths were swapped as Max. was less than Min.!" shown on the screen.

The **Prefix for output filenames** will be applied to the Monte Carlo output files if it is not empty. It is set from the currently open file name with the **Set** button and can be modified. For example, one could add "_1k" to indicate that only 1000 samples are being used. If the prefix is left empty, e.g. if no file has been opened or the set function is not used, the Monte Carlo output files will only be distinguished by their time stamps, and if these are disabled, may overwrite each other.

To keep the program usable on multiple platforms, the file name prefix is restricted to letters, numbers and the underline. If any other characters are used, these will be removed, the preferences window will be switched to the Monte Carlo tab, and a message "Illegal characters removed from Monte Carlo file name prefix!" shown on the screen.

The **Histogram Data** checkboxes determine whether the histogram data is written to files.

B.5.6 Output Tab

The output tab determines the location, contents and naming of the output files.

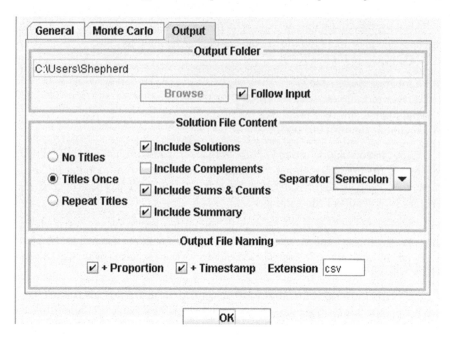

Output Folder

The folder for the output files is preset to the user's home folder.

This can be changed with the **Browse** button, but only if Follow Input is unchecked.

If the checkbox **Follow Input** is checked, the output file will be written to the same folder as the input file that is currently open for file processing. If a new input file is opened, the output folder will "follow" to this new location. Monte Carlo output will also be written to this folder (useful if you want to run a corresponding Monte Carlo after opening or processing an input file).

If the checkbox **Follow Input** is unchecked, the output folder will revert to the one last chosen with Browse (or the home folder if none has been chosen), and the Browse button is activated.

Solution File Content

The contents of the solutions output file can be determined by the checkboxes as follows:

No Titles—the solutions output file will not contain the titles of the pieces.
Titles Once—the solutions output file contains the titles of the pieces from the input file in the first row as headings.
Repeat Titles—the titles are repeated before each solution pair.
Include Solutions—the rows listing the solution pairs are included in the solutions output file.
Include Complements—for each solution row, the complement row is also included with the same index.

Appendix B: Proportional Parallelism Explorer Program User Manual 309

Include Sums and Counts—for each solution row, columns for the sum of the lengths and the number of pieces are included.

Include Summary—the summary is included at the end of the solutions output file—see Sect. B.3.10.

The buttons and checkboxes can be set in any combination—see the following examples:

Titles	Include Solutions	Include Summary	Solutions Output File
No Titles	☐	☐	Empty
	☐	☑	Summary only
	☑	☐	Solutions rows
	☑	☑	Solutions rows and Summary
Titles Once	☑	☑	Titles in first line, Solutions rows, Summary
Repeat Titles	☑	☐	Titles and Solutions (3 rows for each solution)
	☐	☐	Only Title rows (not very useful!)

The above does not apply to the patterns output file—it always includes the titles once and the first solution of the matching pair, and it does not include the summary.

It also does not apply to the Monte Carlo output file—it always includes the titles once and always includes the summary.

For sets with a large number of solutions (e.g. the six cello suites with over 2 billion solutions), the solutions output file can be very large and may even exceed the capacity of the user's disc. For these cases, the "Include Solutions" checkbox can be unchecked to leave only the summary. It is probably the patterns output which is more interesting.

Separator—the separator used between column entries in the output files. For visually examining the output, tab is recommended. (The matrix output always uses tab to keep the display in columns.) For use with Excel, see Sects. B.2.6 and B.5.1. The setting applies to the solutions and patterns output files as well as the Monte Carlo outputs.

Output File Names

The output file names are made up of several components, some of which can be selected in the preferences.

The file processing output files are initially named the same as the input file with an additional "_Solutions" or "_Patterns" for the solutions and patterns output respectively to avoid overwriting the input file, if it has the same extension.

The Monte Carlo simulation output files have "MC_" followed by "Result", "LenHgm" and "ResHgm".

Further components of the name can be added in the options in the output file naming area of the output tab of the preferences. These are:

+Proportion—adds the chosen proportion to the output file name, e.g. "_1to1" for 1:1. "_Opp" is also added if 1:1 with opposites is selected (not applicable to Monte Carlo).

Layers—the layering parameters from the General tab are included next as **_nL** for the number of layers n, and if more than one layer, **S** or **nS** for Strict or not Strict. If strict, this is followed by **S** or **nS** again for Strict proportional count and **A** for all or **F** for only first of the sublayer solutions.

+Timestamp—adds a timestamp to the output file name with year, month, day and hour, minute, second, e.g. "_20171130_153059". The elements are ordered so that the names will sort in chronological order. (Note that in the example below the timestamps for the file processing and Monte Carlo are slightly different because it is impossible to start them both in the same second.)

The combination of proportion and timestamp can be used to avoid overwriting the output files when repeating the run with different inputs or preferences.

Extension—the extension of the output files. This should be chosen as needed for further processing, see Sect. B.5.1.

To keep the program usable on multiple platforms, the extension is restricted to letters, numbers and the underline. If any other characters are used, these will be removed, the preferences window will be switched to the output tab, and a message "File Extension illegal characters removed!" shown on the screen.

The full set of file names from an input filenamed WTC1 with all components, and the Monte Carlo simulation with the prefix set from the input file, appear as follows:

```
WTC1_1to2_2LSnSA_20180118_184715_Solutions.csv
WTC1_1to2_2LSnSA_20180118_184715_Patterns.csv
WTC1_1to2_2LSnSA_20180118_184755_MC_Result.csv
WTC1_1to2_2LSnSA_20180118_184755_MC_LenHgm.csv
WTC1_1to2_2LSnSA_20180118_184755_MC_ResHgm.csv
```

The file names with the minimum components appear as follows:

```
WTC1_Solutions.csv
WTC1_Patterns.csv
MC_Result.csv
MC_LenHgm.csv
MC_ResHgm.csv
```

This is summarised in the diagram below.

Appendix B: Proportional Parallelism Explorer Program User Manual

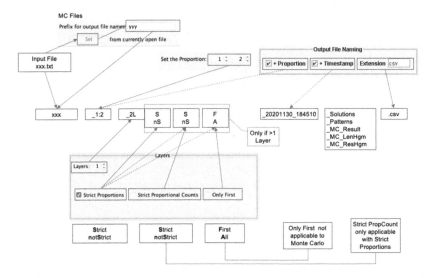

Note: if the program is run using the same output file name as a previous run, the file will be overwritten without any warning. Use the timestamp to avoid this. If the output file is open in another program, e.g. Excel, it cannot be overwritten and an IO[7] Exception error message will appear on the screen.

Note: avoid using extensions reserved for other specific purposes such as .xls, .exe, etc.

B.9 Close

This ends the program. A confirmation is required to avoid accidentally closing it during a run. (Note that no confirmation is required when closing with Command-Q or Quit (on MacOS). Next time you start the program, the preferences will be set to their default values and no input file will be open.

B.10 File Menu

The File menu has:

- Open File
- Start from File
- Pause File
- Resume File
- Cancel File.

---

[7]IO = Input/Output.

B.11 Open File

The input file (see Sect. B.3.3) is selected with the menu item File—Open File. This brings up a file explorer, as usual.

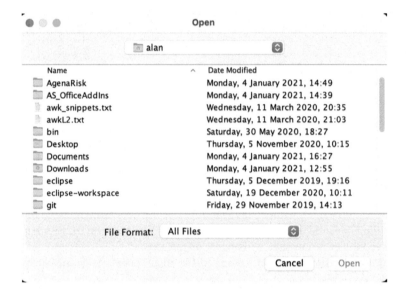

When the program is started, the dialogue shows the user's home folder. Subsequently, it shows the folder last used.

When the file is opened the program reads it and checks it. The errors are shown on the screen, including the line number.

```
Message

Errors in Input file:
Wrong number of words for Piece in line 9
Wrong number of words for Piece in line 10
Invalid length in line 11
Invalid length in line 12
Wrong number of words for Template in line 13
Wrong number of words for Template in line 14
Wrong number of words for Template in line 15
Invalid Action BadCmd in line 16
Invalid Template 012340 in line 17
Wrong number of words for Count in line 18
Wrong number of words for Count in line 19
Invalid count in line 20
Invalid count in line 21
Wrong number of words for BuiltIn in line 22
Wrong number of words for BuiltIn in line 23
Invalid Built In funtion Left=Left in line 24
Invalid Command NewCommand in line 25 - Stopping!
                                                    OK
```

If an invalid command is found, the processing stops at that point. This is to prevent the program running out of memory and crashing if the wrong type of file is accidentally opened.

The path and name of the open file is shown in the main window over the progress bar:

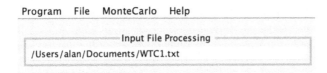

B.12 Start from File

The item "Start from File" in the File menu is disabled until an input file has been selected.

After that, selecting "Start from File" will start the search for solutions and patterns with the opened input file and with the chosen preferences (see Sect. B.8), or with the preset default values if none have been changed.

The process can be run repeatedly with different preferences and/or with different input files.

The input file processing can be started, paused, resumed and cancelled independently of the Monte Carlo processing.

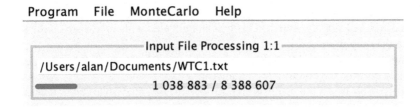

While the search is running, the progress bar is updated in the main window, and the title includes the proportion:

The progress is pessimistic. It is not possible to predict the number of solutions and the 100% mark is the theoretical maximum number of combinations (2^N, or $2^{N/2}$ for 1:1 without opposites). Since the program does not continue to try sequences after the target has been exceeded, it will usually finish quicker than indicated.

For input file processing the progress bar is updated every second and the numerical progress counts the layer 1 tries. For multi-layer runs, this may appear to stop while lower layers are being processed.

When the processing is started, the program also checks that the input file has not been altered since it was opened. If it has, it is read again, and this may produce new error messages.

B.13 Pause File

This pauses the file processing. The remaining part of the progress bar is shown in yellow.

From this state, you can resume or cancel the file processing.

Note: on Mac the colour does not appear.[8] To check if the processing is paused, see if the resume item in the File menu is enabled.

This is useful if you wish to perform other work on the computer and the processing is making the computer too slow.

B.14 Resume File

This resumes the paused processing.

B.15 Cancel File

The Cancel File menu item is available while an input file is being processed or is paused and will stop the search. This will be indicated with a message on the screen and in the output file.

The remaining part of the progress bar is shown in red.

Note: on Mac the colour does not appear.

This can be used if the search is taking too long and you wish to restart it with other preference settings (e.g. with matrix output to observe progress—see Sect. B.8).

A message is displayed with the summary—see Sect. B.3.10.

The output files are produced with the results so far.

B.16 Monte Carlo Menu

The MonteCarlo menu has:

- Start Monte Carlo

---

[8]This feature is not implemented in Java for MacOS.

- Pause Monte Carlo
- Resume Monte Carlo
- Cancel Monte Carlo.

B.17 Start Monte Carlo

Selecting this starts the Monte Carlo simulation with the parameters given in the preferences.

While the simulation is running, the progress bar is updated in the main window.

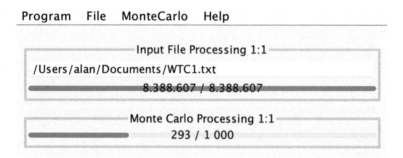

The simulation can be started and cancelled independently of the file processing.

B.18 Pause Monte Carlo

This pauses the Monte Carlo simulation. The remaining part of the progress bar is shown in yellow (not on Mac).

From this state, you can resume or cancel the Monte Carlo simulation.

This is useful if you wish to perform other work on the computer and the processing is making the computer too slow.

B.19 Resume Monte Carlo

This resumes the paused simulation.

B.20 Cancel Monte Carlo

The Cancel Monte Carlo item is available while the simulation is running or is paused and will stop the simulation. This will be indicated with a message on the screen and in the output file. The remaining part of the progress bar is shown in red (not on Mac).

This can be used if the simulation is taking too long and you wish to restart it with other preference settings (e.g. fewer samples).

A message is displayed with the summar—see Sect. B.3.10.

The output files are produced with the results so far.

B.21 Help Menu

The Help menu has:

- Release Notes
- About
- Help
- Extract Manual.

Release Notes shows the version history of the program with the changes made to each version.

About gives the origin and development information of the program.

Help gives assistance with extracting the manual.

Extract Manual will extract this user manual document from the .jar file and open it. It first tries to extract it to the directory in which the .jar file is located, and if that fails, tries to the user's home directory. The location is shown in a confirmation message. The manual will then be opened in the PDF reader. If any of these operations fails, an appropriate message is given—see Help for alternatives.

B.22 Use Cases

B.23 Try Various Proportions

If you wish to generate various proportions for the same input file, set the output file preferences to include the proportion but not the timestamp.

Set the proportion and run the program (click "Start from File" in the File menu) for all desired proportions. The output files will be differentiated by the proportion in their names.

Appendix B: Proportional Parallelism Explorer Program User Manual 317

B.24 Try Various Inputs

If you wish to alter the input file, e.g. by adjusting the lengths of the pieces, set the output file preferences to include the Timestamp.

You can repeatedly edit and save the input file and run the program. The output files will be differentiated by the timestamp in the file names and you will not risk overwriting one with the next.

Alternatively, use different names for the input files to indicate their purpose. This will be reflected in the output file names and the Timestamp is not necessary.

B.25 Performance

Note that each additional piece doubles the number of combinations and so potentially doubles the time the program will take to run to completion. The time also depends on the number of ways the target can be reached with the given lengths as well as the processor speed of the user's computer.

It would be desirable to show an estimate of the time remaining with the progress bars in the main window, but from the nature of the search, which skips a number of combinations, it is clear that a useful estimate cannot be obtained. The situation for Monte Carlo simulation is similar, as it skips combinations which cannot give a solution, and also depends on the solution search algorithm.

The file processing and Monte Carlo simulation can be run simultaneously. Since they run in separate threads, they will use separate processors if your computer has multiple processor cores. Multiple instances of the program can also be run simultaneously to work on different input files or run with different parameters (see Sect. B.2.3). Even if you do not have enough processor cores for these to run faster, they can all be left overnight or over several days to perform all the tasks without further intervention.

B.26 Large Files

Some output files are too large to open in Excel. The following techniques can be used to access them.

Split the file

Split the file into a number of smaller files, e.g. of one million lines each. On MacOS the command is:

```
split -l 1000000 <file>.csv
```

On Windows there is no native command, so either install the Windows Subsystem for Unix to use the above command, or use a third-party utility or script.

Note that the file must be split by lines rather than file size so that lines are not split between files.

Search the file

To find a particular line use a search facility, e.g. on MacOS to find the pattern in layer 1:

```
grep -c "1;Left=Right" <file>.csv
```

Appendix C
Tabular History

The following list (Table C.1) shows some of the most important historical dates and gives an impression of the major figures who have contributed to this field in relation to Bach's lifetime. The material on statistics and mathematics in general is mainly from (Bernstein1996; Cowles1982; Gullberg1997; Knobloch1974; Felbick2012; Göncz 2013; Tatlow 2006).

Table C.1 Brief history of statistics

Year	Event	References
3500 BC	Astralagus (knuckle) bones used as dice for games of chance (or "hazard"—Arabic "al zahr"= dice) in Egyptian tomb paintings and Greek vases	7.2
~500 BC	Pythagoras tuning by mathematical division of string (monochord)	8.1
1st C BC	Vitruvius: Ten books of Architecture	8.1
33 AD	Lots cast for Christ's robe	
13th C	Ramon Lull—symbol permutations Ars compendiosa inveniendi veritatem, 1274 Ars demonstrative, 1283 Ars magna generalis ultima, 1305–8	5.1
1452	Leon Battista Alberti: De re aedificatoria (published 1485)	
1494	Paccioli: Summa di arithmetica, geometria et proportionalità Summary of arithmetic, geometry, proportions and proportionality - includes the problem of dividing the stakes when a game of chance is stopped before its conclusion	Chap. 8
1509	Pacioli. De Divina Proportione - on the divine proportion	
1556	Daniele Barbaro: Italian translation of Vitruvius	8.1
1570	Andrea Palladio: Quattro Libri dell'Architectura—the four books of architecture	8.1

(continued)

Table C.1 (continued)

Year	Event	References
1623	Galileo: Sopra Scoperte dei Dadi On a Discovery Concerning Dice	7.2
1654	Correspondence between de Méré, Pascaland Fermat, including Paccioli's problem	7.2
1657	Huygens: De ratiociniis in ludo aleae On reasoning in games of dice	7.2
1662	Graunt: statistics on mortality etc., sampling	7.2
1663	Cardano: Liber de Ludo Aleae Book of Dice Games published (written in 1525)	7.2
1666	Leibniz: Dissertatio de arte combinatoria Dissertation on the Art of Combination (published in Leipzig!). Includes Partitioning, Combinationsand Permutations as well as probability and related subjects applied to games	5.1, 8.2.3
1679	Leibnizdevelops binary notation, including religious connotations such as $1 = $ God and $0 = $ Nothing, $111 = 7$ days of creation	2.3
1685	*J. S. Bach born in Eisenach*	
1696	Lloyds list—insurance for shipping	
1710	Christian Wolff: Anfangsgründe aller Mathematischen Wissenschaften. Basic Principles of all mathematical sciences	8.1, 13.1
1711	De Moivre: De Mensura Sortis On the Measurement of Lots	7.2
1713	Jacques Bernoulli: Ars conjectandi The Art of Conjecture. Law of large numbers. Coloured pebbles in 3:2 proportion	7.2
1718	De Moivre: Doctrines of Chances (further editions in 1738 and 1756). Binomial approximation, normal distribution, standard deviation	7.2
1723	*J. S. Bach takes the position of Thomaskantor in Leipzig*	
1733	De Moivre: equation for normal distribution	7.2
1730	Euler: Tentamen novae theoriae musicae ex certissimis harmoniae principis dilucide expositae. An attempt at a new theory of music, exposed in all clearness according to the most well-founded principles of harmony	13.1
1737	Mizler: Anfangsgründe aller musikalischen Wissenschaften nach mathematischer Lehrart abgehandelt. Basic principles of all musical sciences treated according to mathematicalteaching methods	13.1
1739	Mizler: Anfangsgründe des Generalbaß nach mathematischer Lehrart abgehandelt. Basic principles of figured bass treated according to mathematicalteaching methods	13.1
1741, 1746	Mizlerpublishes reviews of Euler	
1748	Mattheson: Aristoxeni iunior Phthongologia Systematica. Versuch einer systematischen Klang-Lehre. Essay on a systematic theory of sound	7.3, 13.1

(continued)

Table C.1 (continued)

Year	Event	References
1750	*J. S. Bach died in Leipzig*	
1764	Bayes: Essay Towards Solving A Problem In The Doctrine of Chances published posthumously (written in the 1740s)	7.3, 7.6
1809	Laplace: Central Limit Theorem	7.1, 7.2
	Gauss: Least Squares, Normal (Gaussian) Distribution	7.1, 7.2
1818	Bessel: probable error	
1842–1856	David Ramsay (Hay 1838): a series of books on harmony, proportions, geometry, etc	3.17.4, 8.1
1849	Quetelet: frequencydistribution, application to sociology, concept of "l'homme moyen" the average man	8.1, 7.2
1875	Galton: regression to the mean	
1893	Pearson: fitting observeddistributions to theoretical curves. 1900 Chi squared test of goodness of fit, correlation coefficient, statistical significance	7.3
1906	Gosset: small samples, t-distribution	
1910	Woodand Stratton: significance, standard error. 30 to 1 as lowest odds for practical certainty that a difference is significant Fisher: Significance levels for discrepancy between data and null hypothesis for manageable tables as 0.05	7.3
1912–1913	Whiteheadand Russelpublish Principia Mathematica	2.8, 13.1
1928	Hartley—measure of information	2.2
1930s–1940s	Ulam, von Neumann: Monte Carlo simulationmethods	7.5
1931	Gödel—incompleteness theorem	2.8
1937	Neyman: Confidence interval	7.4
1940	Le Corbusier—Le Modulor	8.1
1948	Shannon—information theory	2.1–2.4
1962	Alberoni: subjective probability, threshold of 0.05 for dismissal that an effect is due to chance	7.3

Appendix D
Alphabet Tables

D.1 Numeric Alphabets

Table D.1 G-values of letters

Letter	Latin Natural	Latin Milesian 1	Latin Trigonal 1
a	1	1	1
b	2	2	3
c	3	3	6
d	4	4	10
e	5	5	15
f	6	6	21
g	7	7	28
h	8	8	36
i	9	9	45
j	9	9	45
k	10	10	55
l	11	20	66
m	12	30	78
n	13	40	91
o	14	50	105
p	15	60	120
q	16	70	136
r	17	80	153
s	18	90	171
t	19	100	190
u	20	200	210
v	20	200	210
w	21	300	231
x	22	400	253
y	23	500	276
z	24	600	300

Letter	Latin Natural	Latin Milesian 1	Latin Trigonal 1
ä	6	6	16
ö	19	55	120
ü	25	205	225
ß	36	180	342
é	5	5	15
'	0	0	0
,	0	0	0
-	0	0	0
.	0	0	0
;	0	0	0
î	9	9	45
á	1	1	1
â	1	1	1
à	1	1	1
ã	1	1	1
ç	3	3	6
è	5	5	15
ë	5	5	15
í	9	9	45
ý	23	500	276
ó	14	50	105
Ò	14	50	105
ú	20	200	210
Ù	20	200	210
ñ	13	40	91

Notes:
"i" and "j" and "u" and "v" are given the same G-Value.
The German Umlauts ä, ö, ü and the ß are treated as ae, oe, ue and ss, as is normal practice.

D.2 G-Values for Notes with Sharps and Flats

Table D.2 G-values of notes

Note (English)	Note (German)	Latin Natural	Latin Milesian 1	Latin Trigonal 1
C♭♭	Ceses	49	193	378
C♭	Ces	26	98	192
C	C	3	3	6
C♯	Cis	30	102	222
C♯♯	Cisis	57	201	438
D♭♭	Deses	50	194	382
D♭	Des	27	99	196
D	D	4	4	10
D♯	Dis	31	103	226
E♭♭	Eses	46	190	372
E♭	Es	23	95	186
E	E	5	5	15
E♯	Eis	32	104	231
E♯♯	Eisis	59	203	447
F♭♭	Feses	52	196	393
F♭	Fes	29	101	207
F	F	6	6	21
F♯	Fis	33	105	237
F♯♯	Fisis	60	204	453
G♭♭	Geses	53	197	400
G♭	Ges	30	102	214
G	G	7	7	28
G♯	Gis	34	106	244
G♯♯	Gisis	61	205	460
A♭♭	Ases	42	186	358
A♭	As	19	91	172
A	A	1	1	1
A♯	Ais	28	100	217
A♯♯	Aisis	55	199	433
B♭♭	Bes	25	97	189
B	B	2	2	3
B♭	H	8	8	36
B♯	His	35	107	252
B♯♯	Hisis	62	206	468

Appendix E
Interval Proportions

Table E.1 Musical intervals as proportions

m:n	Sum	Name (Martineau 2017)	Name (Duffin 2007)
1:1	2	Prime	
2:1	3	Oktave	8ve
3:2	5	reine Quinte	Perfect 5th
4:3	7	reine Quarte	Perfect 4th
5:3	8	reine große Sexte	Major 6th
5:4	9	reine große Terz	Major Third
6:5	11	reine kleine Terz	Minor Third
7:4	11	harmonische Septime	
7:5	12	septimaler Tritonus	
7:6	13	septimale kleine Terz	
8:5	13	diatonische kl. Sexte	Minor 6th
9:5	14	diatonische kl. Septime	Minor 7th (greater)
8:7	15	septimale große Sekunde	
9:8	17	große Sekunde	Major Tone
10:9	19	kleine große Sekunde	Minor Tone
15:8	23	diatonische gr. Septime	Major 7th
16:9	25	pythagoreische kl. Septime	Minor 7th (lesser)
16:15	31	gr. diatonischer Halbton	Major Semitone
27:16	43	pythagoreische gr. Sexte	
25:24	49	kl. diatonischeer Halbton	
27:25	52	großes Limma	

(continued)

Table E.1 (continued)

m:n	Sum	Name (Martineau 2017)	Name (Duffin 2007)
32:25	57		Diminished 4th
32:27	59	pythagoreisches kleine Terz	
45:32	77	diatonischer Tritonus	Augmented 4th
64:45	109		Diminished 5th
75:64	139		Augmented 2nd
81:64	145	pythagoreische gr. Terz	
81:80	161	syntonisches Komma	
128:81	209	pythagoreische kl. Sexte	
125:96	221		Augmented 3rd
128:125	253	Diesis	
135:128	263	gr. chromatischer Halbton	Minor Semitone
225:128	353		Augmented 6th
243:128	371	pythagoreische gr. Septime	
256:135	391		Dimished 8ve
256:225	481		Dimished 3rd
256:243	499	Limma, pythagoreischer Halbton	
405:256	661		Augmented 5th
729:512	1241	pythagoreischer Tritonus	
2025:1024	3049		Augmented 7th
2048:2025	4073	Diaschisma	
2187:2048	4235	Apotome	
32,805:32,768	65,573	Schisma	
531,441:524,288	1,055,729	pythagoreisches Komma	

Appendix F
Excel Functions

To make these functions available to all Excel workbooks, create a new workbook (file) named for example lcb.xlam and copy them into VisualBasic code modules of this file. Then load this file as an add-in. Search for "User Defined Function" or "Excel Add-In" for further information.

An Add-In with these functions and the alphabet tables in Appendix D can be downloaded from https://github.com/Goldberg53/PPExp with no guarantees that it will work as expected.

To edit alphabet tables:

- With any workbook, and the Add-In activated
- In the Developer menu, click Visual Basic
- In the Project, This Workbook set the property "IsAddIn" to False. Set back to "True" when finished.

F.1 Factors of a Number

```
Option Explicit

Public Function Factors(x As Double) As String
' Find factors and prime factors of x
 Dim i As Double
 Dim PrimeFactors As String
 Factors = 1
 PrimeFactors = 1
 For i = 2 To x
 ' If x/i is an integer, then i is a factor
 If x / i = Int(x / i) Then
    Factors = Factors & ", " & i
    If IsPrimeV(i) Then
        PrimeFactors = PrimeFactors & ", " & i
    End If
 End If
 Next i
 Factors = Factors & " Prime: " & PrimeFactors
End Function

Function IsPrimeV(Num As Double) As Boolean
' Test if Num is prime
    Dim i As Double
    ' 2 is even but prime
    If Num = 2 Then
        IsPrimeV = True
        Exit Function
    End If
    ' If Num is even, then not prime
    If Int(Num / 2) = (Num / 2) Then
        Exit Function
    Else
        ' If Num divisible by anything below its square root, then not prime
        For i = 3 To Sqr(Num) Step 2
            If Int(Num / i) = (Num / i) Then
                Exit Function
            End If
        Next i
    End If
    IsPrimeV = True
End Function
```

F.2 G-Value

```
Public Function G_Value(G_String As String, Alphabet As String) As
Integer
' Calculate the gemiatric value (G-Value) of a string using a given
table of alphabet to number conversions.
' V1    May-2016       Alan Shepherd    First Version
' V2    07-Dec-2016    Alan Shepherd    Copy table to an array for
better performance
'                                       Error handling if letter not
found in table
' V3    May-2020       Alan Shepherd    Reworked as Add-In
' Arguments:
'   G_String:    an alphanumeric string to be converted
'   Alphabet     a string denoting the alphabet to be used: L, M, T
for Latin, Milesian or Trigonal
'                These are in the Spreadsheet worksheets of the Add-In

Dim Length As Integer         ' Length of input string
Dim i As Integer              ' Counter through input string
Dim Letter As String          ' Current letter in input string
Dim G_Letter As Integer       ' G-Value of letter (looked up)
Dim G_Array() As Variant      ' Internal copy of alphabet table
Dim Li As Long                ' Counter through array
Dim G_Table As Range          ' Alphabet table

Application.Volatile              ' Volatile to show errors on
calculate even when nothing changed
Application.ScreenUpdating = False    ' Stop updating screen for extra
speed

'Get the range for the selected Alphabet table
Select Case Alphabet
Case "L"
    Set G_Table =
ThisWorkbook.Worksheets("Latin_Natural").Range("A:B")
Case "M"
    Set G_Table =
ThisWorkbook.Worksheets("Latin_Milesian1").Range("A:B")
Case "T"
    Set G_Table = ThisWorkbook.Worksheets("Trigonal1").Range("A:B")
Case Else
    Application.ScreenUpdating = True       ' Resume updating screen
    Dummy = MsgBox("Alphabet " + Alphabet + " Sheet does not exist in
G_Value_Function Add-In", vbOKOnly, "Warning")
    G_Value = -99
    Exit Function
End Select

' Set internal array dimensions to match alphabet table
ReDim G_Array(1 To G_Table.Rows.Count, 1 To G_Table.Columns.Count)

' Copy alphabet table to array for better performance
G_Array = G_Table
```

```
Length = Len(G_String)           ' Length of input string
G_Value = 0                      ' Initialise final result - G-Value of input string

For i = 1 To Length              ' Loop through input string
    Letter = LCase(Mid(G_String, i, 1))         ' Get current letter from input string
    G_Letter = -1                                ' Initialise G-Value of current letter
                                                 ' (not 0, as this can occur as a value)
    Li = 1                                       ' Initialise current letter counter in array
    While G_Letter = -1 And Li <= UBound(G_Array)  ' Loop until letter found or end of array
        If Letter = G_Array(Li, 1) Then          ' Check if the letter matches this array entry
            G_Letter = G_Array(Li, 2)            ' If so, get the G-Value
        End If
        Li = Li + 1                              ' Next letter in array
    Wend

    If G_Letter = -1 Then                        ' If letter not found, give warning
        Dummy = MsgBox("Letter " + Letter + " in word " + G_String + _
" in cell " + Application.Caller.Address + _
        " not in table!", vbOKOnly, "Warning")
    Else
        G_Value = G_Value + G_Letter             ' Add G-Value of valid letter to sum of string
    End If
Next                                             ' Next letter of input string

Application.ScreenUpdating = True                ' Resume updating screen

End Function
```

Literature

Antognazza MR (2016) Leibniz: a Very Short Introduction. Oxford University Press
Bar-Natan D, McKay B (1997) Equidistant Letter Sequences in Tolstoy's "War and Peace". https://users.cecs.anu.edu.au/~bdm/dilugim/WNP/main_en.html. Preprint Accessed 21 Sept 2018
Neumann S et al (ed) Bach Dokumente. Bärenreiter
Bayreuther R (2013) Bach – Mattheson. Zwei deutsche Komponisten und ihre Bücher. In: Damm H, Thimann M, Brill CZ (eds) The artist as reader. On Education and Non-education of Early Modern Artists
Bernstein PL (1996) Against the Gods—the Remarkable Story of Risk. Wiley
Blondel F (1698) Cours D'Architecture Enseigné Dans L'Academie Royale D'Architecture (Band 1): Ou Sont Expliqvez Les Termes, L'origine & les Principes d'Architecture, & les pratiques des cinq Ordres. Paris
Böß R (2009) Die Ungleich Schwebende Originalstimmung von Johann Sebastian Bach Das Wohl Tem Perierte Clavier. Peter Lang, Frankfurt am Main
Bücking J, Heinrich J (1804) Anweisung zur geheimen Correspondenz systematisch entworfen. Wolfenbüttel
Butt J (1997) A mind unconscious that it is calculating? Bach and the rationalist philosophy of Wolff, Leibniz and Spinoza. In: Butt J (ed) The Cambridge Companion to Bach. Cambridge University Press
Cowles M, Davis C (1982) On the Origins of the .05 Level of Statistical Significance. Am Psychol 37(5): 553–558
Cross J (2003) Composing with numbers, sets, rows and magic squares. In: Fauvel J et al (ed) Music and Mathematics—from pythagoras to fractals. Oxford University Press
Cruz N et al Widening access to Bayesian Problem Solving. Front Psychol 11: 660. https://doi.org/10.3389/fpsyg.2020.00660
Daniel LC (1998) Statistical Significance Testing: A Historical Overview of Misuse and Misinterpretation with Implications for the Editorial Policies of Educational Journals. Res Schools 5(2)
Dieben H Bach's Kunst der Fuge. Caecilia en de Muziek Dez. 1939–Jan. 1940 (Given in (Kramer 2019))
Dieben H (1954) Getallenmystiek bij Bach. In: Musica sacra: tijdschrift voor kerkmuziek, vol 5 (S. 21–23) and vol 6 1955 (S47–49). Stichting Lutherse Werkgroep voor Kerkmuziek, Den Haag
Dobelli R (2011) Die Kunst des klaren Denkens – 52 Denkfehler, die Sie besser anderen überlassen. Carl Hanser Verlag, München. Available in English as: The Art of Thinking Clearly
Duffin RW (2007) How Equal Temperament Ruined Harmony (and Why You Should Care). W.W. Norton & Company
Dürr A (2010) Johann Sebastian Bach – Die Kantaten.10. Auflage. Bärenreiter

Fadista J et al (2016) The (in)famous GWAS P-value threshold revisited and updated for low-frequency variants. Eur J Human Gen 24: 1202–1204, Macmillan
Felbick L (2012) Lorenz Christoph Mizler de Kolof – Schüler Bachs und pythagorischer "Apostel der Wolffischen Philosophie". Georg Olms Verlag
Fenton N, Martin N (2013) Risk Assessment and Decision Analysis with Bayesian Networks. CRC Press
Fischer JCF (1935) Ariadne Musica in Liber Organi, Deutsche Meister des 16. und 17. Jahrhunderts Band II. Ernst Kaller. Schott
Gardiner JE (2013) Music in the Castle of Heaven—a Portrait of Johann Sebastian Bach. Penguin Books
Gödel K (1931) Über formal unentscheidbare sätze der "Principia Mathematica" und verwandte Systeme I. Monatshefte für Mathematik und Physik 38: 173–198
Göncz Z (2013) Bach's Testament—on the Philosophical and Theological Background to the Art of Fugue. Scarecrow Press
Grove (2001) The New Grove Dictionary of Music and Musicians, Sadie S, Tyrrell J. (eds). Macmillan
Gullberg J (1997) Mathematics from the Birth of Numbers. Norton
Hadjeres G, Pachet F, Nielsen F (2017) DeepBach: a Steerable Model for Bach Chorales Generation. In: Proceedings of the 34th international conference on machine learning, Sydney, Australia, PMLR 70: 1362–1371. https://arxiv.org/abs/1612.01010
Hardy GH (1940, 1967) A Mathematician's Apology. Cambridge University Press
Hartley RVL (1928) Transmission of Information. Bell Syst Tech J 3: 535–564
Hay DR (1838) The laws of harmonious colouring: adapted to interior decorations, manufactures, and other useful purposes. W. S Orr & Co., London
Hirsch A (1986) Die Zahl im Kantatenwerk Johann Sebastian Bachs. Hänssler, Verlag
Hofstadter DRG (1979) Escher, Bach: an Eternal Golden Braid. Basic Books
Hossenfelder S (2018) Lost in Math, How Beauty Leads Physics Astray. Basic Books
Hubbard DW (2009) The Failure of Risk Management: Why it's Broken and How to Fix It. Wiley
Hubbard DW (2014) How to Measure Anything: Finding the Value of Intangibles in Business, 3rd edn. Wiley
Irwin JL (2015) Foretastes of Heaven in Lutheran Church Music Tradition: Johann Mattheson and Christoph Raupach on Music in Time and Eternity. Rowman & Littlefield
Jones RP (1997) The keyboard works: Bach as teacher and virtuoso in The Cambridge Companion to Bach, Butt J (ed). Cambridge University Press
Kahnemann D (2011) Thinking, Fast and Slow. Farrar, Straus and Giroux
Knobloch E (1974) The Mathematical Studies of G.W. Leibniz on Combinatorics. Historia Mathematica 1
Samuel B (2015) Arrays of Light. Distanz Verlag GmbH, Berlin
Kramer T (2000) Zahlenfiguren im Werk Johann Sebastian Bachs. Dissertation, Utrecht, (reprint 2019)
Klotz S (2014) Kombinatorik und die Verbindungskünste der Zeichen in der Musik zwischen 1630 und 1780. Walter de Gruyter
Leaver RA (1997) The mature vocal works and their theological and liturgical context. In: Butt J (ed) The Cambridge Companion to Bach. Cambridge University Press
Leibniz GW (1666) Dissertatio de arte combinatoria
Leibniz GW (2009) Letter to Philipp Jakob Spener, Hannover, 8(18) Juli 1687. In Gottfried Wilhelm Leibniz Sämtliche Schriften und Briefe, zweite Reihe, zweiter Band. Akademie Verlag
Leibniz GW De Arte Characteristica ad Perficiendas Scientia Ratione Nitentes. In Sämtliche Schriften, Reihe VI Band 4 edn. Academy of Sciences of Berlin. Series I–VIII. Darmstadt, Leipzig, and Berlin
Leibniz GW An Nicolas Remond, Vienna, 10. January 1714, Transkription für die Leibniz-Akademieausgabe der Leibniz-Forschungsstelle Hannover. http://www.gwlb.de/Leibniz/Leibnizarchiv/Veroeffentlichungen/Transkriptionen.htm. Accessed 25th Oct 2018

Translation

Leibniz GW (1989) Letters to Nicolas Remond. In: Loemker LE (eds) Philosophical papers and letters. The new synthese historical library (Texts and Studies in the History of Philosophy), vol 2. Springer, Dordrecht

Licht C (2009) Lied mit geheimer Botschaft – Verschlüsseln von Sprache durch Musik. Junge Wissenschaft No. 83. Verlag Physikalisch-Technische Bundesanstalt. Deutsche Hochschulwerbung und -vertriebs GmbH

Lyons L (2013) Discovering the Significance of 5σ. arXiv:1310.1284v1

Lull R Ars compendiosa inveniendi veritatem, 1274, Ars demonstrative, 1283, Ars magna generalis ultima, 1305–1308

Mäser R (2000) Bach und die drei Temporätsel. Peter Lang, Bern

Marissen M (1995) The Social and Religious Designs of J. S. Bach's Brandenburg Concertos. Princeton University Press

Marissen M (2016) Bach and God. Oxford University Press

Martineau J (ed) (2017) Quadrivium. Librero

Mattheson J (1739) Der vollkommene Capellmeister. Hamburg

Mattheson J (1910) Grundlage einer Ehren-Pforte. Hamburg, 1740. Nachdruck Schneider, Berlin

Mattheson J (1748) Aristoxeni iunior Phthongologia systematica - Versuch einer systematischen Klang-Lehre. Martini, Hamburg

Maul M Bachs Spuren in Weimar – "dem Besitzer zu geneigtem Andenken". Bach Magazin Herbst/Winter 2016/17

McGrayne SB (2011) The Theory That Would Not Die. Yale University Press. Read in the German Edition: Die Theorie, die nicht sterben wollte. Springer Spektrum (2014)

Meredith D Computing Pitch Names in Tonal Music: A Comparative Analysis of Pitch Spelling Algorithms. St. Anne's College, University of Oxford, PhD submission. http://titanmusic.com/papers/public/meredith-dphil-final.pdf. Accessed 27 Aug 2018

Mizler LC (1739) Anfangsgründe des Generalbaß nach mathematischer Lehrart abgehandelt. Leipzig

Ollerenshaw K, Brée D (1998) Most-Perfect Pandiagonal Magic Squares—Their Construction and Enumeration. University Press, Cambridge

Pascha KS (2004) "Gefrorene Musik" Das Verhältnis von Architektur und Musik in der ästhetischen Theorie. Thesis, Faculty of Architecture, Technical University Berlin. https://doi.org/10.14279/depositonce-1036

Pind JL (2012) Looking back: Figure and ground at 100. In: The psychologist, vol 25, No 1, pp 90–91. The British Psychological Society

Prautzsch L (2000) Bibel und Symbol in den Werken Bachs. Thomas-Morus-Bildungswerk, Schwerin

Quantz JJ (1983, 2018) Versuch einer Anweisung, die Flöte traversière zu spielen. Reprint der Ausgabe 1752. Bärenreiter, Kassel

Rowland ID, Howe TN (1999) Vitruvius Ten Books on Architecture. Cambridge University Press

Rubin E (1915) Synsoplevede Figurer: Studier i psykologisk Analyse Første Del. Gyldendalske Boghandel, Nordisk Forlag, Copenhagen

Rumsey D The symbols of the Bach Passacaglia. http://www.davidrumsey.ch. Accessed 28 Feb 2019

Rumsey D (1997) Bach and numerology: 'dry mathematical stuff'? Literature & Aesthetics, Sydney

Sachs K-J (1984) Aspekte der numerischen und tonartlichen Disposition instrumentalmusikalischer Zyklen des ausgehenden 17. und beginnenden 18. Jahrhunderts. Archiv für Musikwissenschaft, 41. Jahrg., H. 4, pp 237–256, Franz Steiner Verlag

Schütz A (1941) Über Zahlenrelation in Präludium und Fuge Es-Dur von Johann Sebastian Bach/S. 27

Schulze H-J (2017) Die Bach-Kantate "Nach dir, Herr, verlanget mich" und ihr Meckbach-Akrostichon. In: Bach-Facetten Essays – Studien – Miszellen. Evangelische Verlagsanstalt, Leipzig, pp 439–449. First published in Bach-Jahrbuch no. 97 in 2011

Schwenter D Harsdörffer, Georg Philipp Deliciae Physico-Mathematicae, Oder Mathemat. und Philosophische Erquickstunden. Nürnberg, Teil 1 (Schwenter) 1636, Teil II (Harsdörffer) 1677, Teil III (Harsdörffer) 1692

Selfridge-Field E Beyond MIDI: The Handbook of Musical Codes. MIT Press (1997)

Shannon CE (1948) A Mathematical Theory of Communication. Bell Syst Tech J 27: 379–423, 623–656

Siegele U (1997) Bach and the domestic politics of electoral saxony. In: Butt J (ed) The Cambridge companion to Bach. Cambridge University Press

Siegele U Johann Sebastian Bach komponiert Zeit – Tempo und Dauer in seiner Musik. Tredition, Hamburg

    Band 1 Grundlegung und Goldberg-Variationen (2014)
    Band 2 Johannes- und Matthäus-Passion (2016)
    Band 3 Wohltemperiertes Klavier I und II (2017)
    Band 4 Tänze und Suiten (2018)

Smend F (1950) Johann Sebastian Bach bei seinem Namen gerufen: eine Noteninschrift und ihre Deutung. Bärenreiter, Kassel

Smend F (1969) Bach-Studien Gesammelte Reden und Aufsätze. Herausgegeben von Christoph Wolff. Bärenreiter, Kassel

Smend F (1966) J. S. Bach Kirchen Kantaten erläutert von Friedrich Smend. Christlicher Zeitschriftenverlag, Berlin

Sonni J, Goodman R (2017) A Mind at Play, The Brilliant Life of Claude Shannon, Inventor of the Information Age. Simon & Schuster

Sterne JAC, Smith GD (2001) Sifting the evidence—what's wrong with significance tests? Brit Med J 322

Sterne J The End of Statistical Significance? Lecture Notes, Department of Social Medicine, University of Bristol

Suchalla E (1994) C.P.E. Bach Briefe und Dokumente. Göttingen

Szabó Z (2016) Problematic Sources, Problematic Transmission: An Outline of the Edition History of the Cello Suites. In: Bach JS (ed) PhD thesis, Sydney Conservatorium of Music, The University of Sydney

Taleb NN (2007) The Black Swan: The Impact of the Highly Improbable. Random House

Taruskin R (2010) Music in the Late Twentieth Century. Oxford University Press

Tatlow R (1991) Bach and the Riddle of the Number Alphabet. Cambridge University Press

Tatlow R (2006) The Use and Abuse of Fibonacci Numbers and the Golden Section in Musicology Today. Understanding Bach 1: 69–85 Bach Network

Tatlow R (2007) Collections, bars and numbers: Analytical coincidence or Bach's design? Understanding Bach 2: 37–58. Bach Network

Tatlow R (2013) Theoretical Hope: A Vision for the Application of Historically Informed Theory. Understanding Bach 8: 33–60. Bach Network

Tatlow R (2015) Bach's Numbers—Compositional Proportion and Significance. Cambridge University Press

Thicknesse P (1772) A Treatise on the Art of Deciphering, and of Writing in Cypher: With an Harmonic Alphabet. London

Thoene H (2016) Johann Sebastian Bach, Ciacconna – Tanz oder tombeau? Dr. Ziethen Verlag

Tomita Y The well tempered Clavier, book 1. Essay from CD liner note of Masaaki Suzuki's performance BIS-CD-831/814. http://www.music.qub.ac.uk/tomita/elinks.html. Accessed 27 Feb 2019

Tomita Y The inventions and sinfonias. Essay from CD liner note of Masaaki Suzuki's performance BIS-CD-100. http://www.music.qub.ac.uk/tomita/elinks.html. Accessed 27 Feb 2019

Walter M (2011) Johann Sebastian Bach Johannespassion Eine musikalisch-theologische Einführung. Carus-Verlag

Walther JG (1732) Musicalisches Lexicon. Wolfgang Deer, Leipzig

Wasserstein RL, Schirm AL, Lazar NA (2019) Moving to a World Beyond "$p < 0.05$". Am Statistician 73: sup1, 1–19. https://doi.org/10.1080/00031305.2019.1583913. Accessed 23 May 2019

Weizenbaum J (1966) ELIZA—a computer program for the study of natural language communication between man and machine. Commun ACM 9(1)
Werker W (1922) Studien über die Symmetrie im Bau der Fugen und die motivische Zusammengehörigkeit der Präludien und Fugen des "Wohltemperierten Klaviers" von Johann Sebastian Bach. Breitkopf & Härtel, Leipzig
Werckmeister A (1707) Musicalische Paradoxal-Discourse der ungemeinen Vorstellungeen. Quedlinburg
Wilkins J (1694) Mercury or the Secret and Swift Messenger. London
Williams P (2001, 2003) Bach: The Goldberg Variations. Cambridge University Press
Wittkower R (1998) Architectural Principles in the Age of Humanism. Academy Editions, (first published 1949)
Wolff C (1710) Anfangsgründe aller Mathematischen Wissenschaften
Wolff C (1991) Bach: Essays on his Life and Music. Harvard
Wolff C (2001) Johann Sebastian Bach The Learned Musician. Oxford University Press
Markus (2006) Dem weltberühmten Manne des Landes Baden ein würdiges Denkmal setzen. Johann Caspar Ferdinand Fischer zum 350. Geburtstag in Musik in Baden-Württemberg, Jahrbuch 2006, Band 13. Strube Verlag

Internet Sources[9]

Internet1: Gosling J et al The Java Language Specification. Oracle America Inc. https://docs.oracle.com/javase/specs/index.html. Accessed 10 Nov 2017
Internet2: Blue Jay Java Development Environment. https://www.bluej.org. Accessed 10 Nov 2017
Internet3: Eclipse Development Environment. https://www.eclipse.org. Accessed 9 Feb 2019
Internet4: GitHub Code Repository. https://github.com. Accessed 4 Feb 2019
Internet5: R Core Team (2018) R: A language and environment for statistical computing. R Foundation for Statistical Computing, Vienna, Austria. https://www.R-project.org/ Accessed 18 Dec 2018
Internet6: Tomita Y Bach, Musicological Font. http://www.music.qub.ac.uk/tomita/bachfont/index.htm. Accessed 29 Mar 2019
Internet7: Musipedia—The Open Music Encyclopedia. http://www.musipedia.org. Accessed 18 Oct 2019
Internet8: Kelley RT Figured Bass Symbols, http://www.robertkelleyphd.com/FiguredBass.pdf. Accessed 1 Dec 2016
Internet9: Ampel FJ, Uzzle T, Clark K The History of Audio and Sound Measurement Pro Audio Encyclopedia http://proaudioencyclopedia.com/the-history-of-audio-and-sound-measurement/. Accessed 5 May 2017
Internet10: Hebblethwaite P 7 secret codes and ciphers hidden in music. https://www.bbc.co.uk/music/articles/cb7ac9cf-207e-4244-8302-2436f2c2ba5a. Accessed 21 Dec 2017
Internet11: Wikipedia. Morse Code. https://en.wikipedia.org/wiki/Morse_code. Accessed 4 Jan 2018
Internet12: Raz C Music of the Squares: David Ramsay Hay and the Reinvention of Pythagorean Aesthetics. https://publicdomainreview.org/2019/05/16/music-of-the-squares-david-ramsay-hay-and-the-reinvention-of-pythagorean-aesthetics/. Accessed 2 Oct 2019

---

[9]Internet web sites come and go and change over time. If any of the references given here have disappeared or have changed, it may be possible to find previous states at The Internet Archive https://web.archive.org.

Internet13: Samuel B (2012) Beethoven und Bach betrachten – Aus Noten werden Bilder. Pressemitteilung 25. http://benjaminsamuel.net/wp-content/themes/bs/pdf/120125_PM_Goldberg-Diabelli_DE.pdf. Accessed 6 June 2020

Internet14: Service T BBC Radio 3 Program The Listening Service. Codes, Ciphers, Enigmas. 10 Sept. 2017. https://www.bbc.co.uk/programmes/b078n25h. Accessed 4 Jan 2018

Internet15: Djossa CA With Musical Cryptography, Composers Can Hide Messages in Their Melodies. Atlas Obscura https://www.atlasobscura.com/articles/musical-cryptography-codes. Accessed 18 Oct 2019

Internet16: Schreiber J Free German Dictionary. https://sourceforge.net/projects/germandict/. Accessed 9 Nov 2016

Internet17: Luther M Luther Bible http://www.gasl.org/refbib/Bibel_Luther_1912.pdf. Accessed on 24 Nov 2016

Internet18: Bischof WF Bach Cantatas Internet, https://webdocs.cs.ualberta.ca/~wfb/bach.html. Accessed 15 Nov 2016

Internet19: List of Baroque Composers https://en.wikipedia.org/wiki/List_of_Baroque_composers. Accessed 15 Nov 2016

Internet20: MIDI Wikipedia, https://en.wikipedia.org/wiki/MIDI. Accessed 13 Nov 2016

Internet21: Selby A asc2mid.exe http://www.archduke.org/midi/index.html. Accessed 13 Nov 2016

Internet22: Haußwald G (ed) (1958) IMSLP 6 Sonatas & Partitas for Solo Violin, VWV 1001–1006. http://imslp.org/wiki/6_Violin_Sonatas_and_Partitas_BWV_1001-1006_(Bach_Johann_Sebastian). Accessed 22 Apr 2017

Internet23: Grossmann DJ http://www.jsbach.net/midi/index.html. Accessed 22 Apr 2017

Internet24: The Humdrum Toolkit: Software for Music Research. https://www.humdrum.org/. Accessed 6 Sept 2019

Internet25: Music Encoding Initiative MEI https://music-encoding.org. Accessed 6 Sept 2019

Internet26: MuseData https://musedata.org. Center for Computer Assisted Research in the Humanities at Stanford University. http://www.wiki.ccarh.org. Accessed 6 Aug 2019, Links updated 29 Apr 2020

Internet27: Music21 http://web.mit.edu/music21. Accessed 7 Sept 2019

Internet28: It's time to talk about ditching statistical significance Editorial 20 March 2019. Nature Int J Sci https://www.nature.com/articles/d41586-019-00874-8. Accessed 23 May 2019

Internet29: NIST's New Quantum Method Generate *Really* Random Numbers. https://www.nist.gov/news-events/news/2018/04/nists-new-quantum-method-generates-really-random-numbers. Accessed 25 May 2019

Internet30: Mehta T (2012) Computer program used as basis for combinations. http://www.tusharmehta.com/excel/templates/match_values/. Accessed 28 Feb 2017

Internet31: Objects of Belief: Proportional Systems in the History of Architecture. Special Issue of The Open Access J Euro Architect Hist Netw Ubiquity Press. June 2014. https://journal.eahn.org/collections/special/objects-of-belief-proportional-systems-in-the-history-of-architecture/. Accessed 7 Aug 2019

Internet32: Wikipedia. List of Cognitive Biases. https://en.wikipedia.org/wiki/List_of_cognitive_biases. Accessed 20 Mar 2017

Internet33: Measuring Perceptions of Uncertainty from Heuer RJ (1999) Psychology of intelligence analysis. CIA Center for the Study of Intelligence. https://www.cia.gov/library/center-for-the-study-of-intelligence/csi-publications/books-and-monographs/psychology-of-intelligence-analysis/. Accessed 21 Aug 2019

Internet34: The Organ of Don and Jill Knuth https://www-cs-faculty.stanford.edu/~knuth/organ.html. Accessed 7 May 2020

Internet35: You'd never know it wasn't Bach (or even human). Yale News, August 20 2015. https://news.yale.edu/2015/08/20/you-d-never-know-it-wasn-t-bach-or-even-human. Accessed 26 Aug 2019

Internet36: Garcia C Algorithmic Music—David Cope and EMI. Computer History Museum. April 29, 2019. https://www.computerhistory.org/atchm/algorithmic-music-david-cope-and-emi/. Accessed 26 Aug 2019

Internet-CC-Licenses: Creative Commons Licenses. https://creativecommons.org/licenses/

General Index

A
Abacus, 23, 39, 226
Accidental, 27, 36, 58
Acrostic, 29–30, 61
Aesthetics, 93, 94, 101, 196, 333, 335
AgenaRisk, xv, 83, 128
Algorithm, 32, 80, 103, 104, 229, 317
Alla breve, 180, 254
Alles mit Gott und nichts ohn' ihn, 23
Alphabet, ix, 3, 6, 8, 11–13, 15, 16, 19, 22, 23, 25, 27, 29, 31, 34, 36, 38, 39, 43, 48, 49, 54, 77, 187, 209, 230, 323
    Latin Natural, 3, 6, 11, 12, 25, 29, 39, 41, 42, 44, 45, 48, 51, 54, 55, 76, 77
    Milesian, 25, 39, 42, 43, 46, 48, 55, 76
    Trigonal, 39, 42–44, 46, 48, 55
Ambiguity, 14, 15, 25, 38, 55, 72
American Statistician, The, 78, 335
Anacrusis, 59, 60, 62–65, 72, 94, 179, 180, 196, 274
Anagram, 15
Ansbach, 219
Antisemitism, 215
Apophenia, 214
Apostles, 16
Appoggiatura, 59
Approximation, 80, 82, 87, 320
Architecture, xv, 24, 32, 88, 90, 180–184, 196, 220, 259, 319
Ariadne Musica, 149–151
Arithmetic, 9, 74, 91, 100, 183, 230, 319
Artificial intelligence, xxiii, 9, 227, 227–230, 231
Assumption, 82, 85, 129–131, 214
Astronomy, 219

Athens, 87
Atom, 221
Average, 42, 44, 45, 48, 53, 74, 76, 77, 112, 124–127, 153, 165, 166, 170, 172, 191, 197–203, 207, 215, 245, 263, 266, 271, 275. *See also* Mean

B
Bach Archive, xv
Bach Dokumente (BD), xxiii, 220, 331
Bach Museum (Eisenach), 32
Bach Museum (Leipzig), 23
Bach Werke Verzeichnis, xxiii, 31
Batch file, 45
Bayesian inference, 129
Bayesian network, xv, xvi, 82, 83, 85, 128–133
Bayes theorem, 73, 74, 82, 85, 128–133, 132
Bell Laboratories, 11
Bias, xiv, 50, 73, 180, 196, 235, 336
    confirmation, 214, 215
    hero, 215
    story, 179, 213–214
Bible, 34, 40, 44, 45, 48, 336
Bible reference, 34
Binary, 8–10, 76, 99, 100, 103, 107, 109, 206, 226, 292, 293, 320
Birthday, 23, 215
Bit, 8–11, 13, 76, 80, 99, 108, 109, 192, 293, 307
Blacksmith, 2, 87
Boson, 222
Brandenburg Concertos, 6, 20, 154, 168, 173, 179, 185, 186, 192, 197, 200, 222, 237–241, 333

Bullseye, 121
BWV 75. *See* Die Elenden sollen essen
BWV 76. *See* Die Himmel erzählen die Ehre Gottes
BWV 77. *See* Du sollt Gott, deinen Herren, lieben
BWV 150. *See* Nach dir, Herr, verlanget mich
BWV 1001-1006. *See* Sei Soli for Violin
BWV 165. *See* O heiliges Geist- und Wasserbad
BWV 176. *See* Es ist ein trotzig und verzagt Ding
BWV 194. *See* Höchsterwünschtes Freudenfest
BWV 232. *See* Mass in B minor
BWV 244. *See* St. Matthew Passion
BWV 245. *See* St. John Passion
BWV 248. *See* Christmas Oratorio
BWV 552. *See* Prelude and Fugue in E major
BWV 772-801. *See* Inventions and Sinfonias
BWV 825-830. *See* Clavierübung I
BWV 831. *See* Clavierübung II
BWV 846-869. *See* Well Tempered Clavier Book 1
BWV 870-893. *See* Well Tempered Clavier Book 2
BWV 971. *See* Clavierübung II
BWV 988. *See* Goldberg Variations
BWV 1007-1012. *See* Cello Suites
BWV 1046-1051. *See* Brandenburg Concertos
BWV 1079. *See* Musical Offering
BWV 1127. *See* Alles mit Gott und nichts ohn' ihn
BWV numbers, 31, 275
Byte, 9, 11, 13, 57

C
Cadence, 32
Calculator, 52, 95, 226, 228, 230, 231
Canon, 20, 222, 224
Canonic Variations, 241–244
Cantata, 6, 24, 26, 45, 46, 48, 219, 336
Capacity, 7, 11, 13, 19, 32, 309
Cello Suites, 20, 154, 168, 174, 185, 186, 190, 192, 194, 200, 222, 244–245, 309, 334
Cent, 87
Central Intelligence Agency (CIA), xxiii, 75, 336
Central limit theorem, 74, 77, 124, 172, 321

Chance, 7, 38, 48, 65, 66, 73, 74, 77, 79, 83, 93, 118, 121, 125, 127–131, 149, 165, 166, 183, 190, 196, 209, 216, 217, 245, 319, 321
Channel, 7–11, 13, 19, 32
Chess, 228
Chorale, 19, 67, 155, 168, 173, 174, 177, 182, 185–187, 198, 201, 222, 228, 241, 248, 259, 261–263, 332
Christmas Oratorio, 24
Chromatic, 32
Ciaccona, 23, 26, 60, 63, 65, 67, 262
Circle of fifths, 222
Clavichord, xiii, 37, 87, 245
Clavierübung I, 6, 24, 26, 154, 168, 173, 185, 192, 198, 200, 245
Clavierübung II, 26, 154, 168, 173, 185, 192, 198, 200, 245–248
Clavierübung III, 248–249
Coding, 39, 40, 42–45, 47, 48, 54, 65, 70, 72, 77, 214, 215
 direct, 16, 25, 30, 49
 indirect, 16, 22, 24–26, 35
 reference, 2, 8, 16, 17, 19, 20, 23, 24, 26, 27, 29, 30, 34, 38, 49, 54, 61, 106, 335
Coincidence, 171, 172, 222, 224, 334
Collection, **6**
Colour, 32, 73, 90
Colour coding, 32–33, 111
Combination, xiv, 26, 27, 31, 32, 34, 37, 52–54, 74, 80, 82, 95–99, 100, **102**, 103–106, 110, 118, 121, 126, 128, 129, 132–134, 136, 138–140, 153, 164–167, 170–172, 183, 191, 197–202, 206, 218, 219, 226, 230, 244, 245, 250, 254, 255, 263, 274, 290, 291, 295, 297, 309, 310, 313, 317, 320, 336
Communications technology, 8, 13
Communication theory, 7–12
Compact disc, xxiii, 179
Complement, 97–103, 99, **102**, 104, 105, 107, 109, 110, 121, 136, 141, 210, 251, 255, 268, 275, 290, 292, 293, 295, 297, 305, 306, 308
Composer, 3, 9–12, 21, 23, 49, 55, 72, 98, 111, 112, 170, 180, 181, 193, 212, 215, 216, 220, 227, 229, 254, 256, 273, 336
Composition, **5, 6**, 13, 20, 23, 24, 26–29, 32, 34–36, 102, 111, 180, 182, 183, 215, 216, 219–221, 224, 228, 230, 231, 235, 244, 262, 263, 273

General Index 341

Compression, 14
Computational musicology. *See* Musicology
Computer, xiv, xv, xxiv, 1–3, 8–11, 13, 14, 19, 22, 23, 37, 38, 40, 52, 57, 59, 71, 74, 80, 93, 99, 100, 102, 103, 109, 186, 196, 205, 217, 220, 226–228, 228–231, 233, 234, 266, 284, 285, 287, 288, 314, 315, 317
Computer aided musicology, 2
Confidence interval, 73, 79
Continuo, 29
Conventions, 5
Counterpoint, 216, 230
Cryptography, 12, 34, 336. *See also* Encryption
Curve fitting, 172

D

Da capo, 60, 94, 117, 177, 179, 183, 196, 223, 244, 245, 247, 259, 261, 262
Dal segno, 60, 244
Dark energy, 222
Dark matter, 222
DARMS, xxiii, 71
Dartboard, 121
Database, 13, 14
Decimal, 9, 99, 109, 288, 289, 301
Decoding, 7, 14–16, 22, 25, 37–39, 39, 51, 54
Decompression, 14
Definitions, 5–6, 93–103
DeoxyriboNucleic Acid (DNA), xxiii, 222
Deutsche Gesellschaft, 219
Diabelli variations, 32
Dice, 74, 212, 230, 319, 320
Dictionary, 1, 40–42, 44, 46, 48, 53, 54, 222
Die Elenden sollen essen, 24
Die Himmel erzählen die Ehre Gottes, 24
Digital Alternate Representation of Musical Scores. *See* DARMS
Dissonance, 87, 136
Distribution, 41, 42, 44, 65, 66, 74, 74–77, 77, 80, 82, 85, 112, 123, 124, 172, 174, 192, 196, 207, 236, 262, 266, 271, 281, 300, 320, 321
  Gaussian. *See* Distribution, Normal
  normal, 41, 74, 76–78, 80, 124, 172, 174, 192, 196, 217, 266, 320
  uniform, 76, 80, 112, 123, 170, 266, 271, 300
Double. *See* Proportion, double
Duration, 6, 22, 36, 37, 58, 182–184, 196, 223

Du sollt Gott, deinen Herren, lieben, 29

E

Economics, 213
Edition, 20, 187, 190, 244, 332–334
Egypt, 74, 319
Eisenach, 32, 320
Electron, 221, 222
ELIZA, 228, 335
Encoding, 7, 39, 43, 47, 49, 57, 70, 71, 76, 233, 336. *See also* Coding
Encryption, 11, 22, 34
Engineering, 8, 220
English Suites, 249–250
Enigma Variations, 33
Entropy, 8, 9, 14
Equal tempered, 26, 222
Equidistant letter sequences, 70, 331
Es ist ein trotzig und verzagt Ding, 26
Estimate, 73, 79, 80, 203, 215, 317
Exact pattern matching, 108, 109, 292, 293
Excel, 15, 16, 39, 41, 52, 53, 58, 95, 98, 103, 106, 107, 110–112, 185, 193, 234, 266, 288, 299, 301, 302, 309, 311, 317, 327
Excel add-in, 327
Explorer. *See* Proportional Parallelism Explorer
Exponential, 153, 164–166, 230, 253
eXtensible Markup Language (XML), xxiv, 71

F

Factor, 94, 95, 196, 281, 305, 328
Factorial, 52
Fallacy, 213, 214, 216
  conjunction, 216
  narrative, 213, 214
Fenton House, xiii
Fermion, 222
Figured bass, 29, 226
File
  input, 104, 136, 289, 298–302, 305–311, 314, 315–317
  output, 103, 115, 289, 298–302, 305–311, 314–317
  processing, 290–298
Filter, 98, 111
Filtering, 103, 110, 185, 301
Finale, 72
Forgery, 229
Formatting, 98, 106, 113, 288, 301

French Overture. *See* Clavierübung II
French Suites, 250–254
Frequency, 8, 9, 31, 32, 34, 65, 67, 70, 74, 112, 113, 172, 191, 222, 300–302, 321, 332
Fretting, 87
Fugue, 6, 11, 20, 23–25, 29, 54, 117, 187, 209, 222, 229, 231, 258
Function, 13, 41, 95, 109, 112, 284, 290, 292, 294, 302, 307, 327
Fuzzy search, 181

G
Gaussian. *See* Distribution, Normal
Gematria, 6, 13, 14, 16, 31, 38, 39, 49, 65, 66, 70, 72, 77, 92, 205, 233
Genetic code, 12, 222
Genetics, 16, 78, 222
Geometry, 3, 74, 219, 319, 321
Go, 228
Gödel number, 15
Goldberg Variations, 6, 24, 28, 32, 154, 166, 168, 173, 184, 191, 198, 201, 223, 254–256, 335
Golden section, 87
Goodness of fit, 74, 321
Gravitational waves, 222
Greatest common divisor, 133, 135
Great Fifteen Organ Preludes, 256
Greece, 12, 15, 32, 39, 74, 94, 221, 319
Grove, 5, 6, 332
G-Value, **6**, 6, 13, 15, 16, 22, 23, 25, 35, 39, 41, 42, 44–46, 48, 51, 52, 55, 58–60, 65, 67–70, 77, 241, 258, 324, 329

H
Halle, 218
Halting problem, 230
Hammer, 2, 87
Hardware, 80, 100, 109
Harmony, 2, 16, 29, 36, 87, 95, 219, 230, 320, 321
Harpsichord, 29, 32, 37, 38, 239, 245, 248, 273, 280
Hash coding, 13–15, 15, 25, 39, 49
Hashing. *See* Hash coding
Hi-Fi. *See* High fidelity
Higgs boson, 78, 221
High fidelity, 230
Histogram, 41, 45, 73, 76, 80, 82, 112, 113, 115, 123, 172, 175, 253, 255, 266, 281, 299–302, 307

Historically Informed Theory, 3, 93, 127, 179, 190, 192, 219, 228, 334
Höchsterwünschtes Freudenfest, 26
Holy Trinity, 16, 24, 26
Horn, 37
Hourglass, 182
Human body, 32, 88, 90, 91, 223
Humdrum, 71, 336
HyperText Markup Language (HTML), xxiii, 45, 71
Hypothesis testing, 70, 73, 77–79, 82, 127, 196

I
ICT. *See* Information and Communications Technology
Impact, 74
Index, 45, 98, 100, 106, 110, 192, 295, 297, 308, 335, 336
Information and Communications Technology, xxiii, 8
Information technology, 8
Information theory, xiv, 3, 76, 233, 321
Input file. *See* File, input
Instrumentation, 30
Insurance, 74, 320
Integer, 87, 94, 102, 183, 291, 293, 307
Interference, 7, 9, 11, 13
Invention, 219, 231
Inventions and Sinfonias, 6, 24, 133, 136, 155, 168, 173, 174, 190, 192, 198, 201, 256–258, 334
Investment, 213
Irrational, 28, 94
Isopsephy, 39
Israel, 215
Italian Concerto. *See* Clavierübung II
Italy, 11, 90, 154, 229, 245, 273, 319
Iteration, 104

J
Java, xv, xxiii, 103, 284, 285, 288, 289, 314, 335
Jazz, 179
Joint Photographic Experts Group (JPEG), xxiii, 14

K
Kern, 71
Key, 25, 26, 32, 37, 87, 91, 92, 133–136, 149, 222

General Index 343

Keyboard Partitas. *See* Clavierübung I
Key signature, 25–26, 26, 36
Köthen, 244, 262, 280

L

Latin Natural Order Alphabet. *See* Alphabet, Natural order
Law, 24, 219
Layer, 91–93, 100–102, **102**, 103, 106, 107, 110–112, 115, 117, 122, 123, 126, 128, 131–133, 136, 140–142, 148, 151, 153, 164–169, 174, 181, 192, 196, 209, 210, 223, 236, 237, 239, 240, 242, 254, 252, 256, 264–266, 268–272, 275, 277–281, 295, 297, 299, 301, 305, 306, 310, 313, 317
Left=Right, 109, 136, 140–142, 148, 151, 184, 190, 235–238, 240, 242–244, 246, 247, 249, 251, 254, 256, 257, 259, 261, 264, 265, 272, 273, 275, 276, 279, 280, 297, 317
Left=thgiR, 109, 110, 136, 140, 142, 151, 184, 190, 235–238, 240, 242–244, 246, 247, 249, 251, 252, 254, 256, 257, 259, 261, 264, 265, 272, 273, 279, 280, 297
Leipzig, xv, 23, 24, 38, 52, 96, 149, 182, 215, 217–220, 258, 274, 280, 320, 320, 321, 332, 333, 334
Le Modulor, 90, 321
Length, **6, 102**
Lepton, 221, 222
Libretto, 180
Likelihood, 8, 75, 76, 80, 82, 83, 124, 129, 131, 149, 167, 171, 174, 209, 215
Linda, 216
Literature, xiv, 219
Lloyds, 74, 320
Logarithm, 8
Logic, 1, 77, 131, 217, 218, 288

M

Machine learning, 228, 229
Magic rectangle, 206, 207, 209, 224, 233
Magic square, 20, 205–212, 224, 331
Manuscript, 10, 11, 20, 127, 180, 196, 235, 244
Margin of error, 79, 80
Mass in B minor, 6, 24, 27, 111, 173, 174, 176, 182, 192, 235–237, 254

Mathematics, 1–3, 15, 16, 34, 51, 52, 74, 82, 91, 93, 95, 96, 109, 131, 205, 213–215, 217–220, 222–224, 230, 231, 233, 319, 320, 333
Matrix, 205, 206, 290, 298, 305, 309, 314
Maximum, 11, 24, 38, 42, 45, 48, 66, 111, 112, 123–127, 172, 207, 266, 271, 298, 299, 303, 305, 307, 313
Mean, 74, 77, 112, 127, 134–136, 172, 249, 299, 321
Mean tone, 136
Medical trials, 78, 128
MEI, xxiii, 71, 336
Message, 1, 7–14, 19–30, 31, 32, 34–36, 38, 48, 57, 65, 70, 76, 213, 214, 216, 285, 286, 289, 293, 304, 305, 307, 310, 311, 313–316
Method, 1, 3, 34, 39, 57, 73, 83, 93, 105, 109, 111, 117, 118, 153, 181, 183, 184, 193, 196, 218, 229, 230, 233, 234, 287, 320, 321, 336. *See also* Quantitative methods
Metronome, 37
MIDI, xxiii, 32, 36, 37, 57–65, 62, 64, 65, 67–71, 334, 336
Milesian alphabet. *See* Alphabet, Milesian
Minimum, 106, 111, 112, 123–127, 155, 165, 166, 172, 266, 298, 299, 307, 310
Mirror image, 100, 109, 110, 136, 190, 235, 236, 251, 294
Model, 83, 131, 213, 220–226, 223, 332
Monad, 221
Monochord, 87, 319
Monte Carlo simulation, 74, 80–81, 85, 103, 111, 115, 123, 126, 127, 129, 153, 167, 172, 174, 175, 179, 185, 187, 194, 203, 236, 237, 239, 243, 245, 247, 249–251, 259, 261, 263, 266, 271, 273, 275, 281, 287, 298, 303, 307, 309, 310, 314, 315, 317, 321
Morse code, 31, 32
Movement, **6**
MP3, xxiii, 14
Muon, 222
Murphy's Law, 215
MuseData, 71, 336
MuseScore, 72
Music21, 71, 336
Musical box, 38
Musical Instrument Museum of the University of Leipzig, 38

Musical Instruments Digital Interface. *See* MIDI
Musical Offering, 6, 30, 92, 155, 168, 174, 185, 186, 192, 198, 201, 258, 258–259, 259
Music Encoding Initiative. *See* MEI
Musicology, xiv, xv, 1–3, 30, 71, 91, 93, 136, 231, 233
MusicXML, 71

N

Nach dir, Herr, verlanget mich, 29, 333
Nature, 78, 336
Neural network, 229
Neutrino, 222
Neutron, 221
Node, 83, 129
Node Probability Table, 83, 129
Noise, 7, 9–11, 13, 22, 180, 214
Normal distribution. *See* Distribution, Normal
North Atlantic Treaty Organisation (NATO), xxiii, 75
NP-complete, xxiii, 230
Nucleus, 221
Null hypothesis, 77, 79, 127, 128, 321
Numerologist, 39, 205

O

Observation, 73, 77, 79, 82, 129, 130, 153, 221, 223, 227, 321
Octave, 21, 22, 25, 31, 34, 36, 58, 87, 95, 222
Odds, 74
O heiliges Geist- und Wasserbad, 26
Opposite, 3, 77–79, 96, 98, **102**, 104, 105, 109, 117, 136, 172, 293, 295, 297, 305, 310, 313
Optimisation, 105, 106, 111, 250, 297
Order, 3, 6, 8, 12–15, 22, 29, 42, 51, 52, 54, 55, 98, 104, 106
Organ, xiii, 11, 21, 29, 37, 182, 222, 227, 229, 248, 273, 336
Orthotonophonium, 222
Outlier, 153, 192
Output file. *See* File, Output
Overtones, 37

P

Painting, 196, 229, 319
Palindrome, 100, 109, 235
Paragram, 3, 13, 26, 38, 39, 217
Parallel proportion. *See* Proportional Parallelism
Parameters, 103, 107, 112, 123, 290, 298, 301, 304, 307, 310, 314, 317
Parity, 11, 76
Particle physics, 78, 128, 191, 221, 222
Partition, **51**, 52–55, 94, 96, 320
Pattern, 67, 93, 99, 100, 102, **103**, 105, 106, 107–111, 111, 136–149, 181, 190, 193, 196, 206, 214, 220, 235–237, 239, 241, 244, 245, 248, 250–252, 256, 264, 271, 273, 275, 276, 290–294, 296, 297, 306, 309, 313, 317
Percentile, 112, **125**, 126, 149, 174–176, 178, 179, 187, 196–203, 236, 237, 239, 247, 249, 256, 259, 262, 263, 273, 275, 281
Perfect number, 222
Performance
  musical, 6, 37, 57, 59, 71, 182, 184, 231, 256
  speed, 317
Permutation, 26, 51, **52**, 54, 74, 76, 96, 205, 206, 217, 219, 230, 231, 319, 320
Philosophy, 3, 5, 216, 217–220, 229, 331
Phrase, **6**
Phrygian, 149, 151
Pi, 15, 87, 94
Piano roll, 37, 38
Piece, **5, 102**
Pitch, 23, 31, 36, 37
Pitfall, xiv, 2, 58, 73, 79, 82, 233
Planet, 209
Plotting, 76, 112, 113, 299–302
Poetry, 39, 217, 219
Poland, 218, 235
Population, 73, 74, 79, 80, 82, 126
Prejudice, 216
Prelude and Fugue in E major, 24
Prime factor, 16
Prime number, 15, 16, 184, 217, 259
Principia Mathematica, 15, 218, 321, 332
Prior, 82, 85, 129–131
Probability, xiv, 2, 8, 30, 73, 74, 74–77, 77–80, 82, 83, 85, 93, 118, 121, 123, 125, 128–133, 136, 149, 153, 164–167, 170–172, 190, 191, 196–203, 209, 214–217, 226, 254, 263, 274, 320, 321

General Index 345

Program, xiv–xvi, xxiii, 1, 2, 14, 40, 52, 53, 58, 71, 72, 80, 83, 93, 98, 102, 103–115, 117, 122–124, 129, 136, 141, 170, 183, 186, 194, 206, 220, 226, 228–230, 234, 236, 237, 244, 245, 256, 268, 271, 274, 275, 283–317, 297, 335, 336. *See also* Proportional Parallelism Explorer
PropCount. *See* Proportional Count
Proportion, **102**
   architectural, xv, 32, 90, 180–184
   double, 91, 94, **102**, 103, 107, 110, 141, 155, 164, 167, 170, 194, 236, 237, 245, 256, 263, 264, 275, 280, 281
   Proportional Count, **102**, 106, 110, 117, 119, 136, 141, 155, 164, 192, 295, 297, 305
   strict, **103**
Proportional Parallelism, 91, 93, 94, 103, 110, 117, 136, 141, 193, 196, 209, 229, 233, 237, 239, 245, 248, 256, 263, 268, 275, 281, 283
Proportional Parallelism Explorer, xv, 93, 103, 103–115, 136, 153, 167, 170, 181, 183, 185, 187, 192, 206, 209, 226, 229, 254, 266, 268, 283–317
Protocol, 13
Proton, 221
Pseudo-random number, 80, 170, 298, 300
Psychological fallacies, 213–216
Psychoanalyis, 228
Pulpit, 182
P-value, 73, 77, 79

Q

Quantitative methods, 1, 2
Quantum, 80, 226, 336
Quark, 221, 222

R

R, xv, 103, 112, 113, 299, 335
Random collection, 112, 124, 125, 128, 129, 185, 249, 273, 275
Random number, 77, 80, 111, 171, 300
Random sample, 73, 74, 77, 80, 83, 111, 112, 123, 127, 149, 169, 170, 174, 185, 196, 263, 271, 298
Range, xiv, 7, 28, 42, 43, 65, 75, 79, 80, 82, 123, 126, 128, 153, 155, 166, 167, 170, 172, 191, 196, 248, 266, 300
Ratio, 16, 87, 94, 190, 196, 213, 222
Rational, 87, 213

Recitative, 20
Reconstruction, 9, 187, 190, 226
Rectangle, 205, 209, 233
Recursion, 104, 107
Redundancy, 9–11, 13, 14, 22, 30, 76
Reference. *See* Coding, Reference
Reflection, 100, 109, 193. *See also* Left=thgiR
Regression, 74
Repeat, 13, 20, 22, 59–65, 72, 94, 117, 169, 173, 174, 177–179, 183, 185, 186, 191, 196, 199, 202, 223, 239, 242–248, 259, 261, 273–277, 279–281, 289
Repetition, 27
Rest, 27, 28, 60, 235
Reverse engineering, 186–188
Ricercar, 30, 259
Risk, xiv, 74, 83
Rome, 74, 88
Rounding, 183
Rubin's vase, 27

S

Sample, 9, 44, 54, 58, 73, 74, 77, 79, 80, 82, 85, 93, 104, 111, 112, 123, 125–128, 149, 153, 170, 172, 174, 175, 196, 203, 215, 236, 245, 253, 255, 256, 263, 266, 271, 298–303, 307, 315, 320, 321. *See also* Random Sample
Sample set, 111, 112, 127, 128, 298, 299
Sand-glass, 182
Savart wheel, 31
Scale
   measurement, 8, 49, 253
   musical, 22, 29, 31, 32, 34, 67, 87, 133, 219, 222, 230
Schübler Chorales, 259–262
SCORE, 72
Sculpture, 196
Section, **5**, 6, 20, 23, 24, 111, 179
Sei Soli for Violin, 20, 25, 26, 58, 65, 70, 167, 194, 262–272
Semibreve, 22, 154, 168, 173, 180, 185, 186, 191, 196, 197, 200, 235–238
Semitone, 67, 87
Sequence, 8, 11, 13, 15, 21–23, 29, 31, 76, 80, 88, 100, 103, 104, 111, 136, 205, 207, 266, 271, 293, 295, 297
Sequence numbers, 13, 104
Sermon, 182, 183
Set. *See* Sample Set, Symbol Set

Shift, 108–110, 293, 297
Signal, 9, 10, 180
Signature, 25, 26, 28, 99, 100, **103**, 106–109, 206, 226, 293
Significance. *See* Statistical Significance
Sinfonias. *See* Inventions and Sinfonias
Slide rule, 226
Sociology, 74, 78, 128, 321
Software, 57, 71, 220, 284
Soli Deo Gloria, 205
Solution, 97, **102**
Solution pair, 99–102, **102**, 104, 106, 110, 117, 118, 121, 264, 290, 292, 295, 297, 305, 308
Sorting, 44, 45, 103, 106, 111, 192, 250, 301, 310
Source code, 186
Spectrum, 32
Spreadsheet, 53, 98
Standard deviation, 77, 112, 127, 217, 299, 320
State, 7, 129, 314, 315
Statistical significance, 16, 21, 29, 31, 73, 74, 77–79, 85, 127, 128, 191, 213, 215, 321, 331, 334, 336
Statistics, xiv, xv, 1–3, 48, 65, 73, 74, 80, 103, 153, 196, 197, 203, 214, 217, 231, 233, 235, 299, 319, 320
Stile antico, 180
St. John Passion, 23, 24
St. Matthew Passion, 20, 24, 256
Storage, 7, 11, 13, 226
Strict, 101, **103**, 107, 110, 115, 123, 126, 128, 136, 140–142, 151, 164, 166, 192, 237, 238, 240, 242, 268, 272, 275, 279, 280, 297, 301, 305, 306, 310
St. Thomas' Church, 182
St. Thomas school, 219
Stufendynamik, 37
Sudoku, 207
Summary, 107, 112, 291, 296, 297, 299, 301, 305, 309, 314
Symbol, 8–38, 76, 77, 110, 226, 319
Symbol set, 8–38, 222
Symmetry, 93, 98–100, 103, 109, 133, 136, 141, 149, 181–183, 187, 193, 196, 224, 235, 236, 239, 241, 245, 248, 256, 264, 266, 271, 273, 275. *See also* Left=Right and Left=thgiR
Synthesiser, xiii, 57

T
Tail, 41, 172
Tangent, 87
Target, 98, **102**, 104–107, 118, 121, 206, 254, 297, 305, 313, 317
Tau, 222
Temperament, 31, 36, 37, 54, 87, 136, 220, 222
Template, 100, **103**, 107–110, 136, 220, 290, 292, 293, 297, 336
Tempo, 5, 37, 183, 222
Ten commandments, 16, 24
Terminology. *See* Definitions
Test data, 106
Text editor, 45, 58, 291, 301
Theology, 24, 91, 196, 219, 332
Thermodynamics, 8
Threshold, 2, 53, 54, 77, 78, 128, 191, 245, 259, 321, 332
Timbre, 37
Time signature, 6, 21, 28, 37, 60, 72, 117, 180, 183, 196, 248
Tolerance, 180, 181, 184
Tool, 1–3, 58, 71, 83, 93, 107, 167, 233
Toolkit, 2, 71, 336
Total length, **102**
Transcribed Concertos, 273–274
Translation, 109, 193, 235
Transmission, 7, 8, 10, 11, 13, 30, 32
Transposition, 67, 98, 192, 206, 212
Trigonal alphabet. *See* Alphabet, Trigonal
Trill, 59
Trinity. *See* Holy Trinity
Trio Sonatas for Organ, 274–281
Trumpet, 23, 29, 37
Truncating, 183
Tuning, 26, 31, 36, 37, 54, 136, 220, 222, 319
Tuning fork, 31

U
Uncertainty, 8, 73, 76, 83
Uniform distribution. *See* Distribution, Uniform
Unison, 87, 91, 95
Universe, 205, 213, 222, 231
User manual, 93, 103, 110, 283–317

V
Violin, 20, 23, 26, 36, 37, 70, 93, 167, 239, 262, 280
Violin solo. *See* Sei Soli for Violin

Violin Sonatas, 194, 280–281
Vitruvian man, 88, 223
Vocabulary, 44

W
Weimar, 54, 219, 244, 273, 333
Well Tempered Clavier Book 1, xxiv, 6, 20, 23, 24, 26, 29, 80, 91–95, 100, 112, 149, 153, 155, 169, 173, 181, 184, 190–192, 199, 202, 205, 207, 209, 224, 229, 235, 281, 292, 310, 334
Well Tempered Clavier Book 2, xxiv, 26, 141
Wolf dissonance, 136
Work, **6**
WTC. *See* Well Tempered Clavier

Z
ZIP, xxiv, 14

Index of Names

A
Adelung, Johann Christoph, 40
Alberoni, Francesco, 321
Alberti, Leon Battista, 319
Altnickol, Johann Christoph, 187, 250
Antognazza, Maria Rosa, 217, 331
August III, Elector of Saxony, 235

B
Babbage, Charles, 226
Bach, Anna Magdalena, 244, 245
Bach, Carl Philipp Emanuel, 187, 220, 230, 258, 334
Bach, Maria Barbara, 65, 67, 68
Bach, Wilhelm Friedemann, 55, 136
Badia, Carlo Agostino, 55
Barbaro, Daniele, 90, 319
Bar-Natan, Dror, 70, 331
Bayes, Thomas, xv, 73, 74, 82, 85, 128, 132, 321
Bayreuther, Rainer, 220, 331
Beethoven, Ludwig van, 13, 32, 65, 67, 69, 256, 336
Benevoli, Orazio, 55
Bernoulli, Jacques, 320
Bernstein, Peter L., 73, 319, 331
Bessel, Friedrich, 321
Blondel, François, 180, 331
Böß, Reinhard, 26, 31, 331
Bohr, Niels, 221
Boulez, Pierre, 212
Box, George, 224
Brahms, Johannes, 21
Brixi, Šimon, 55

Bücking, Johann Josef Heinrich, 34, 331
Busoni, Ferrucio, 229
Butt, John, 190, 220, 230, 239, 331, 332, 334

C
Cardano, Gerolamo, 320
Carlos, Wendy, xiii
Casals, Pablo, 244
Castel, Louis Bertrand, 32, 90
Chadwick, James, 221
Cowles, Michael, 77, 319, 331
Crick, Francis, 222
Cross, Jonathan, 212, 331
Cruz, Nicole, 83

D
Daniel, Larry C., 73, 78, 331
Darwin, Charles, 222
Davis, Peter Maxwell, 212
Diabelli, Anton, 32
Dieben, Henk, 3, 20, 205, 209, 224, 331
Dobelli, Rolf, 213, 331
Duffin, Ross W., 87, 325, 326, 331
Dürr, Alfred, 26, 30, 40, 331

E
Elgar, Edward, 33
Escher, Maurits Cornelis, 16, 332
Euler, Leonhard, 219, 320

F
Faber, Johann Christoph, 23

Index of Names

Fadista, João, 78, 332
Felbick, Lutz, 218, 319, 332
Fenton, Norman, xiii, xiv, xvi, 73, 79, 81, 82–84, 215, 332
Fermat, Pierre de, 320
Ferrari, Benedetto, 55
Fibonacci (Bonacci, Leonardo), 88
Fischer, Johann Caspar Ferdinand, 149, 168, 197, 199, 203, 331, 332, 335
Fisher, Ronald Aylmer, 321
Forkel, Johann Nikolaus, 220, 274
Frederick the Great, Friedrich II, King of Prussia, 258
Frisius, Johannes, 40
Fritsch, Thomas, 40

G

Galilei, Galileo, 320
Galton, Francis, 321
Gardiner, John Eliot, 23, 24, 332
Gauss, Carl Friedrich, 41, 74, 76, 80, 172, 192, 321
Geminiani, Francesco, 55
Gesner, Matthias, 219
Gödel, Kurt, 15, 16, 47, 217, 321, 332
Goethe, Johann Wolfgang von, 32, 90
Goldberg, Johann Gottlieb, 254
Göncz, Zoltán, 187–189, 217, 219, 319, 332
Goodman, Rob, 11, 334
Gosset, William Sealy, 78, 321
Gottsched, Johann Christoph, 219
Graunt, John, 74, 320
Grimm, Jacob and Wilhelm, 40
Gullberg, Jan, 73, 319, 332

H

Hadjeres, Gaëtan, 228, 332
Händel, Georg Friedrich, 31
Hardy, G. H., 223, 332
Harsdörffer, Georg Philipp, 96, 226, 334
Hartley, R. V. L., 8, 11, 321, 332
Hay, David Ramsay, 32, 33, 90, 321, 332, 335
Haydn, Michael, 34
Higgs, Peter, 78, 221
Hirsch, Arthur, 20, 23, 332
Hofstadter, Douglas R., 12, 16, 332
Hossenfelder, Sabine, 213, 223, 332
Hubbard, Douglas W., xiv, 76, 332
Huygens, Christiaan, 320

I

Irwin, Joyce L., 78, 332

J

Jacomelli, Geminiano, 55
Jones, Richard P., 332

K

Kahnemann, Daniel, 213, 214, 216, 332
Kayser, Bernhard Christian, 249
Kent, Sherman, 75
Keyserlingk, Count Hermann Carl von, 254
Kircher, Athanasius, 21, 32, 87, 209
Kirnberger, Johann Philipp, 37, 222, 230
Klotz, Sebastian, 32, 38, 230, 332
Knobloch, Eberhard, 319, 332
Knuth, Donald, 227, 336
Koren, Benjamin Samuel, 32, 332
Korsyn, Kevin, 190
Kramer, Thijs, 20, 22, 39, 87, 141, 142, 205, 206, 331, 332

L

Laplace, Pierre-Simon, 74, 321
Leaver, Robin A., 181, 332
Le Corbusier, Charles-Édouard Jeanneret, 90, 321
Leibniz, Gottfried Wilhelm, 2, 52, 96, 214, 215, 217–221, 226, 229, 320, 331–333
Leonardo da Vinci, 88, 229
Licht, Christiane, 34, 35, 333
Lovelace, Ada, 225
Lull, Ramon, 52, 215, 216, 224, 317, 333
Luther, Martin, 40, 44, 45, 48, 336
Lyons, Louis, 78, 333

M

Malcolm, George, 229
Marissen, Michael, 29, 30, 333
Martineau, John, 325, 326, 333
Mäser, Rolf, 20, 205, 224, 225, 333
Mattheson, Johann, 78, 219, 220, 320, 331–333
Maul, Michael, 55, 333
McGrayne, Sharon Bertsch, 82, 333
Meckenbach, Conrad, 29
Mendelssohn-Bartholdy, Felix, 182, 256
Méré, Antoine Gombaud Chevalier de, 320
Meredith, David, 59, 333
Mersenne, Marin, 230

Metius, Adriaan, 87
Mizler, Lorenz Christoph, 183, 216–219, 226, 241, 320, 332, 333
Moivre, Abraham de, 320
Moog, Robert, xiii
Moroney, Davitt, 190
Mozart, Wolfgang Amadeus, 34, 65, 69, 230, 256

N
Neumann, John von, 321
Newton, Isaac, 32
Neyman, Jerzy, 321

O
Ollerenshaw, Kathleen, 205, 333

P
Paccioli, Fra Luca Bartolomeo de, 74, 319, 320
Palladio, Andrea, 90, 319
Pascal, Blaise, 320
Pascha, Khaled Saleh, 91, 333
Pasqualini, Marc'Antonio, 55
Pearson, Karl, 78, 321
Pind, Jörgen L., 221, 333
Porta, Giovanni, 21
Praetorius, Michael, 183
Prautzsch, Ludwig, 20, 34, 333
Pythagoras, 2, 32, 87, 94, 218, 219, 222, 319, 331, 335

Q
Quantz, Johann Joachim, 57, 58, 87, 88, 230–231, 231, 258, 333
Quetelet, Adolphe, 74, 321

R
Rainer, Arnulf, 229
Raphael, 87
Remond, Nicolas, 214, 332, 333
Reusner, Esaias, 55
Roman, Johan Helmich, 55
Rowland, Ingrid D., 88, 90, 333
Rubin, Edgar, 27, 333
Rumsey, David, 20, 22, 43, 333
Russell, Bertrand, 15, 218, 321
Rutherford, Ernest, 221

S
Sachs, Klaus-Jürgen, 134, 135, 222, 333
Schott, Gaspar, 21
Schübler, Johann Georg, 155, 259
Schulze, Hans-Joachim, 30, 331, 333
Schumann, Robert, 21
Schütz, Adalbert, 24, 333
Schwenter, Daniel, 21, 96, 334
Selfridge-Field, Eleanor, 72, 334
Shannon, Claude, 8–11, 34, 321, 334
Shore, John, 31
Shostakovich, Dmitri, 21
Siegele, Ulrich, 182–184, 223, 225, 334
Smend, Friedrich, xiv, 3, 4, 17, 19, 24, 27, 42, 48, 181, 334
Sonni, Jimmy, 11, 334
Spinoza, Baruch, 230, 331
Sterne, Jonathan A. C., 73, 78, 334
Stratton, Frederick John Marrian, 78, 321
Suchalla, Ernst, 220, 334
Suppig, Friedrich, 222
Szabó, Zoltán, 244, 334

T
Taleb, Nassim Nicholas, 213, 214, 334
Taruskin, Richard, 212, 334
Tatlow, Ruth, ix–xi, xv, 2–4, 12, 13, 17, 20, 21, 23, 26, 28, 29, 34, 39, 40, 42, 60–64, 88, 91–95, 98, 117, 133, 167, 170–172, 180–184, 187–190, 192–194, 196, 209, 215, 217, 219, 220, 222, 223, 229–231, 241, 263, 264, 334
Telemann, Georg Philipp, 32
Thicknesse, Philip, 34, 334
Thoene, Helga, 4, 23, 25, 58, 60–65, 67, 70, 334
Thomson, Joseph, 221
Tolstoy, Leo, 70
Tomita, Yo, xvi, 24, 334, 335
Torvalds, Linus, xxiii
Tukey, John, 8
Tversky, Amos, 213, 214, 216

U
Ulam, Stanislaw, 321

V
Vallotti, Francesco Antonio, 55
Vierdanck, Johann, 55
Vitruvius, 88, 90, 319, 333

Vivaldi, Antonio, 34, 53, 273, 274

W
Wagner, Richard, 215
Walter, Meinrad, xiii, 229, 332, 334, 336
Walther, Johann Gottfried, 55, 222, 334
Wasserstein, Ronald L., 78, 334
Watson, James, 222
Weizenbaum, Joseph, 228, 335
Werckmeister, Andreas, 16, 37, 222, 335
Werker, Wilhelm, 3, 335
Whitehead, Alfred North, 15, 218, 321

Wilhelm Ernst, Duke of Sachsen-Weimar, 23
Wilkins, John, 34, 335
Williams, Peter, 28, 335
Wittkower, Rudolf, 90, 196, 335
Wolff, Christian, 91, 217–220, 320, 331, 332, 335
Wolff, Christoph, 181, 187–189, 219, 335
Wood, T. B., 78, 321

Z
Zepf, Markus, 149, 335

Printed by Printforce, the Netherlands